CW01497329

The Gardener's Book of
Weeds

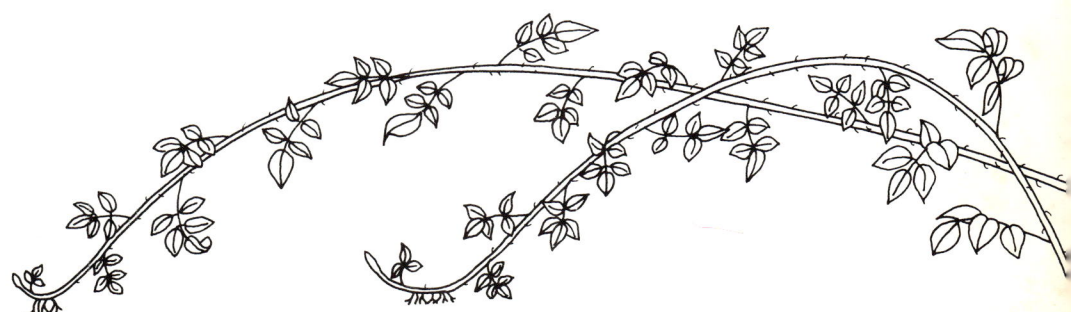

Other books by Mea Allan

Fiction
LONELY
CHANGE OF HEART
ROSE COTTAGE
BASE RUMOUR

Biography
THE TRADESCANTS
THE HOOKERS OF KEW
TOM'S WEEDS
PALGRAVE OF ARABIA
E. A. BOWLES AND HIS GARDEN
PLANTS THAT CHANGED OUR GARDENS
DARWIN AND HIS FLOWERS

FISONS GUIDE TO GARDENS
GARDENS OF EAST ANGLIA

Weeds

The Unbidden Guests in Our Gardens

Mea Allan

With drawings by Victoria Matthews

THE VIKING PRESS · NEW YORK

All the photographs in the book were specially taken by
Grace Woodbridge with the exception of that on
page 40 which is by Heather Angel.
The drawings on page 29 are by Vicky Fisher.

Copyright © Mea Allan, 1978

All rights reserved

Published in 1978 by The Viking Press
625 Madison Avenue, New York, N.Y. 10022

Published simultaneously in Canada by
Penguin Books Canada Limited

Library of Congress Cataloging in Publication Data

Allan, Mea.
  Weeds.

    1. Weeds.   I. Title.
SB611.A45   635.9'2'58   77-17484
ISBN 0-670-75657-1

Designed and produced by
Walter Parrish International Ltd, London
Designer Roger Hyde

Printed and bound in Great Britain by
Purnell & Sons Ltd

Contents

THIS BOOK is a selection of the commonest weeds afflicting European (including British) and North American gardens, a selection arrived at by means of a questionnaire circulated to gardeners and horticultural authorities all over the United States, Britain and continental Europe. Care has been taken to arrange the weeds in a way easily understandable to the home gardener.

They are described alphabetically under the English-language name of their group such as Geraniums, Clovers, and so on; and because gardeners may find members of a group confusingly alike the author, aided by illustrative drawings, has concentrated on pointing out the differences between them.

The scientific (Latin) names of the British and continental European weeds in the book accord with the nomenclature used by Clapham, Tutin and Warburg (*Flora of the British Isles*, 2nd edition) who in general followed the *Check List of British Vascular Plants* issued by the British Ecological Society in 1946. For purely American weeds the author has adopted both the scientific and common names used in Gray's *Manual of Botany* (8th edition) and by the Agricultural Research Service in *Selected Weeds of the United States*, which are generally those of the Weed Science Society of America. But as most of the American weeds are introduced, for those originating in Europe the author has again followed the nomenclature of Clapham, Tutin and Warburg. In the case of weeds introduced into Britain from the United States where they are natives, priority is given to the American names.

Author's Introduction

I can claim some acquaintance with weeds. Following World War II, with anti-invasion defences still littering the sand and shingle beaches of eastern England, my father bought a garden — the cottage that went with it was to him incidental, though its wartime occupation by the Military explained the state of the garden.

It was an acre of weeds. The enclosed rose garden might have been the garden of the Sleeping Beauty: entrance to it between yew hedges was barred by a man-high tangle of prickly briers. Opposite, leading from another cut between yew hedges, the sunken garden was carpeted and entwined by Bindweed. In the long border there, running below a pebble wall, Crinum Lilies were all but smothered in Ground Elder. The vegetable garden was choked with grasses and patches of well-grown nettles. Brambles cut off whole sections of the garden: cascading in impenetrable curtains from trees, they had rooted and climbed up again over bushes. Not a path was visible: weedy ways merged with border, and border with overgrown hedge. Ponds were choked with duckweed and the heaving pads of water lilies. The pleached walk of Huntingdon Elm had a 12ft top growth of thick-as-my-arm upright branches.

Yet, 'What a beautiful garden,' my father said. For every brier was jewelled with sweet-smelling Dog Roses, and every Bindweed with pink-and-white trumpets spiralling round the stiff stems of Flax flowers that cast over that little sunken garden a dreaming haze of blue. In every border the tall spires of Foxgloves — white, pink and darker pink — stood proud as soldiers, and the whole place was a sea of curly-spurred Columbines from palest pink to blue and lustrous maroon. Even the Ground Elder was showing its best in umbels of starry white flowers. The grasses in the kitchen garden, whenever a breeze stirred them, rolled and rippled like the waves of the sea.

The moral of this story is that there is beauty in weeds, though we were all agreed that if we were to have a proper garden it could not remain like this. A flame-gun seemed the quick answer. But as the previous owner had been an artist who was also a plantsman, and locally the garden was still called 'the Lily Garden', we knew that underneath the strangling, tangling mantle of green must be some rare and precious plants. So annihilation of this sort was out of the question: we would have to uncover with care for what might be discovered.

The task of slaughter and rescue began — and it took more than a decade before strength was restored to roses weakened by suckering, before the white strings of Couch Grass ceased their underground tunnellings, before broken dinner knives had prised the last limpet-like knob of Ground Elder from between paving stones, and the last devourer of sunlight and soil nutrient was consigned to the bonfire to make a good nitrogenous beauty treatment for the plants we wished to grow. It was impossible to save the weeds for composting: this would have meant mountainous

Trumpet call of Bindweed

heaps all over the place, and time was not on our side. Our motto was burn and start afresh.

Of course it did not stop there. 'One year's seeding — seven years' weeding', we knew, and there had been seven years of *seeding*. I forgot to mention the ivy. The apex of the long triangular garden, a dense shade under Holm Oaks, was carpeted thick with it, and hardly a tree had escaped its embrace. Shifted on a pitch-fork, pile by pile as we tore it down, the bonfire crackled like a merry Guy Fawkes Night.

In the nearly thirty years I have lived with the garden I have come to know weeds intimately. Some I have come to hate, and some I have come to love. To all I have given respect, even if only in the pitting of my lesser strength against their greater, their determination, their inventiveness, their almost appalling fecundity. Through weeds I have learned a lot about life, the very basics of Darwinian struggle for existence and triumph of survival — though, ultimately, not at my hands. Away with rank Ground Elder, I have cried, for all its starry prime; away with forests of stinking Garlic Mustard; away with Bindweed that snarls my precious alpines. But stay, with a gentler touch, enough of the little bronze-leaved Sorrel to make golden patches among the Suffolk pamments, and stay the wild Geraniums — the seldom-found Dusky Cranesbill (which E.A. Bowles called Venus's Navel) for the black-purple beauty of its flowers, the pink charm of Herb Robert, the miracle of veining in the petals of *versicolor*; nor would I sacrifice the bright berries, shiny green and red, of *Arum maculatum*, and the fascination of watching Diptera flies emerge dusted with pollen from imprisonment.

Some that have no merit at all in the way of beauty I let live for usefulness: such as the chickweed that grows between the pea-sticks. Our soil dries up in summer to dust, and chickweed keeps it moist. Also, peas hate having their roots disturbed, which would happen if I pulled up the straggling chickweed. As to the little Scarlet Pimpernel it visits the garden sparingly, and it amuses me to have this tiny clock to tell me by its open or closed flower whether it is morning or afternoon, though as a Poor Man's Weather-glass it is not nearly so reliable.

They have their stories. Volumes have been written on the legends connected with weeds and wildflowers. Even the humble Daisy had its moment of glory. In the Middle Ages in the tilting yard

> In his scarf the knight the Daisy bound,
> And dames at tourney shone with Daisies crown'd.

They are, too, an aspect of history. The story of the introduction of plants into Britain and America tells of the invasions of civilizing man. To England with the Romans came among other plants the Corn Cockle, Field Spurrey and Rough-headed Poppy: with the monks came culinary and medicinal herbs — and the Borage escaped from the *herbularis* to make waysides gay with its blue rocket flowers. To America with the Pilgrim Fathers came the Dandelion, Shepherd's Purse, Stinging Nettle, Couch Grass, Chickweed, Groundsel and ubiquitous Plantain.

Weeds can be benevolent or toxic. *Rhus radicans*, the Poison Ivy, is dangerous to humans, and even the latex from the broken stem of the Petty Spurge can burn fingers uncomfortably. Some, like the Dodder, are parasitic, living on other plants, or partially parasitic like the Yellow Rattle which sucks water from the roots of the grasses amongst which it grows. Others excrete harmful substances which depress the growth of neighbouring plants, and some become transit hotels for pests which overwinter in them. In temperate areas we say of a mild winter, when in the following summer our garden is plagued by greenfly, that it has not 'killed off the insects'. Much truer to say that it has not killed off the weeds that harbour them. It is well to be able to recognize such plants if they arrive in our gardens.

So there is much to know about weeds. And if they are capable of rousing the gardener's wrath and making him or her despair of ever winning the battle, let me assure him and her there are ways of eradicating them or at least controlling them. Some ways are safe. Some are dangerous, and across the world the alarm bells have sounded. Man, with all the materials of death in his hands, can apply extinction when and where he will. He is being warned that his life and the lives of animals, insects and plants are interdependant. This is not to say that when he uproots a Dandelion in his garden he weighs the scales of survival against himself. But certainly he is contributing to this when he uses a certain type of aerosol. The propellant in the innocent-looking spray releases fluorocarbons which are damaging the earth's ozone-shield, and its use has now been banned in America.

On the other side, a new study is revealing that some weeds may actually be worth encouraging for the help they give to garden plants, both by benefiting their growth and by quelling or killing the pests that attack them, and even by killing more pernicious weeds.

Each of the subjects outlined in this Introduction is treated separately in the book, so that you can make more of your garden by making the most of your garden's weeds.

Grow ornamental grasses in your garden – but don't let them turn into weeds!

1

The World of Weeds

What Is a Weed?

The question of what is a weed and what a wildflower has been variously answered. A weed is 'a plant out of place,' declared Professor W.J. Beal, weed specialist at Michigan State University. 'A plant that grows so luxuriantly or plentifully that it chokes out all other plants that possess more valuable nutritive properties,' said W.E. Brenchley, another weed expert. The *Oxford English Dictionary* defines a weed as 'a herbaceous plant not valued for use or beauty, growing wild and rank, and regarded as cumbering the ground or hindering the growth of superior vegetation.' R.M. Moore's definition is 'a plant which interferes with man's utilization of land for a specific purpose'. Lawrence J. King, who wrote a monograph on the subject, drew up a table listing ten characteristics commonly associated with weeds. Some were incontestable. 'Ability to thrive almost anywhere' was one that wrung grim acquiescence from my gardening heart, as did 'large populations' and 'survival of plant fragments; bio-chemical resistance'. But there have been kinder viewpoints, including L.H. Bailey's 'Nature ... knows no plants as weeds.'

I have known gardeners regard as a weed anything they did not actually sow or plant themselves. I have seen with horror the blank spaces caused by the removal and consignment to a bonfire of rare plants that were merely unfamiliar. To this type of gardener 'A plant growing where it is not desired' (the Terminology Committee of the Weed Society of America) is adequate definition.

The American poet Emerson was nearer the truth with his classic remark that a weed is 'a plant whose virtues have not yet been discovered'. For today all sorts of virtues are being discovered in weeds. We know now that some weeds, in fact most of them, can do positive good: they may stimulate or help to feed their neighbours, excrete insect-repelling odours, destroy harmful bacteria. The Parsley tribe was hitherto regarded as a gang of robbers, their deep roots stealing from the soil great quantities of minerals and nutrients required by other plants; but bio-dynamic studies have revealed that by bringing up these substances which are too deep down for other plants to reach, and storing them in their stems and leaves, the Parsleys can ultimately benefit the garden. By burning the deep-grubbers, the minerals they have absorbed can be given to the topsoil in the ashes. Deep-rooted weeds can also open up the soil for the roots of other plants, allowing them to forage farther. So, far from being worthless and damaging, they are much more mother-plants and soil-builders. Apart from burning, most weeds, even Couch Grass, can be composted if sufficient heat is generated, and the good they have absorbed given back to the soil.

Of course, they are not all beneficent. Buttercups, for instance, secrete a poisonous substance in their roots which acts as a depressant on other plants. Many overwintering weeds are hosts for fungal and bacterial plant diseases. The ethylene gas exhaled by the Dandelion inhibits the height or growth of its neighbours: but on the other hand, the Dandelion is one of the best subsoilers! The host weeds should be burned. But what to do with the Dandelion?

We shall be studying the potentials of weeds good and bad, with their special influences on other plants. Then we shall know what a weed really is, and see them with more understanding.

Where Do Weeds Come From?

Weeds have travelled about the world, and still do travel.

Dispersal is mainly by seeds, and one has only to look at a collection of seeds to know some of the ways they go on their journeys. Take, for instance, the seed of Enchanter's Nightshade: the hooks on its fruit will readily catch on to the fur of animals, to be shaken off at a distant spot. Cleavers will similarly cleave. Plantain and Chickweed seeds, which become sticky when wet, may also adhere to a passer-by and so be dispersed. Note that burred and sticky seeds always belong to low-growing plants. Succulent fruits, too, depend both on animals and birds for their dispersal: their bright colours or shininess are beacons to attract them, and such seeds are able to pass through their digestive tracts without coming to harm, indeed germinate much sooner for having been acted upon by the gastric juices and warmth of the animal's body. Squirrels collect acorns and store them in underground larders. If some of the hoard is not eaten, a forest of oak seedlings will later spring up. Ants store seeds as nesting material, and in this way Gorse seeds are buried and grow up in a new place.

Birds are probably by far the most important agents of dispersal. Look at a bird migration map and see how far the swallow flies from Europe to its wintering areas in central and South Africa — a distance of 5,000–7,000 miles; the bobolink from the North American meadowlands flying 5,000 miles to Bolivia, Paraguay or Brazil; the record-holding Arctic Tern flying from the north of Greenland to the edge of the Antarctic pack-ice — 11,000 miles. The feet of birds often have little cakes of earth attached to them. Alfred Newton, Professor of Zoology at

Poppies and a dock on a day in June. Two months earlier this garden was well tended and hardly a weed was to be seen. Each of the poppies eventually shed 20,000 seeds, the dock about 30,000. Weeds are past masters at the takeover business.

Cambridge, sent Darwin the leg of a partridge with a hard ball of earth weighing $6\frac{1}{2}$ oz adhering to it. Darwin kept the earth for three years, but when he broke it up, watered it and placed it under a bell-glass, no fewer than 82 plants grew from it.

Even fish may help to disperse seeds. An autumn gale may break off the branch of a riverside plant on which there are seeds, and a fish may snatch at them; and this fish or another may be caught by an angling heron, a bird that ranges widely in search of such prey. And so the seed of this weed may be ejected at some other riverside or mud-flat.

Charles Darwin did many experiments on the length of time seeds remain viable after immersion in sea-water. Having been naturalist to the *Beagle*, a surveying ship, he knew the average rate of the sea-currents, and on the reckoning of the average rate of the Atlantic current being 33 miles a day, seeds which survived even 14 days' immersion could be transported 300 miles or even more. Other seeds survived 21 days' immersion. The survival record was won by fresh seed of the Wild Cabbage which germinated excellently after 50 days, very well after 100 days, while two seeds germinated after 133 days. Seeds he took from the crop of a dead pigeon that was floated in his sea-water tank for 30 days nearly all germinated.

The wind is the dispersal agent for winged and plumed fruits and seeds. How perfect for flying are the aeroplanes of the Sycamore, the parachutes of the thistle and Dandelion! Caught up in a gust of wind a plumed seed can travel many miles, and it is interesting that the hairs of the plume spread out only when the air is dry and fit for flying, and close up vertically when it is damp — though this, too, is a way of dispersal, for in rainy weather they fall to the ground and are washed into the soil or sail away in a rivulet. There is a song about the 'Tumbling Tumbleweed', the Russian Thistle. By the time its seeds are ripe, it has rolled itself into a huge and prickly ball and only awaits a gust of wind to break off from the root and be bowled for miles and miles, scattering seed as it goes.

Some weeds use ballistics to disperse their seeds. The Hairy Bitter Cress (*Cardamine hirsuta*) is one. Tiny as its pods are you can hear them exploding on a sunny day, to shoot the seeds to a distance of up to 32in (80cm). Vetches, Gorse and Broom bushes do the same thing. The fruits of the cranesbills split into five portions, each coiling up like a watch-spring so abruptly that the seeds are ejected over several yards.

Icebergs are sometimes loaded with earth and stones. Logs and other driftwood sail across the seas and fetch up on foreign beaches. Tucked into interstices, cracks and crannies are sure to be some seeds.

Can we wonder that, finding new homes in a like or better climate, weeds have travelled far?

How Weeds Came to Britain

Anciently when a land bridge joined Europe and Britain their floras were identical. They were still joined in the Tertiary period, and it was only after a series of upheavals and subsidences with the sea covering and re-covering large parts of the continent, and retreating, that they were finally separated. Then came the glacial periods, four of which afflicted Britain. Plants growing from seeds blown to the lee of the icy breath survived to scatter their own seeds, and so the trek continued, ever southward. In a continent like America, the plants could flee in this way till they reached the tropics, where the glaciers could not reach them. British plants could go no farther than Cornwall. A few lived on, inhabiting nunataks, geologically havens by-passed by the ice-sheets, such as Ben Lawers in Scotland and The Burren in Ireland. But these were alpine plants, accustomed to the cold. Apart from others in a non-glaciated strip between the Thames and the Severn and the English Channel, and a small area in South Wales, Britain was devoid of vegetation.

So we must take as our starting-point the end of the last Ice Age when the British

Isles began to enjoy a moist oceanic climate favourable for invasion — first by plants from the Mediterranean. Northern continental Europe was also losing its last ice-sheet, and in due time, when the plants had spread back again to their old homes, it too sent its emigrants across the Channel. Once more the two floras became identical, or almost so.

The plants became established, and centuries later when man started cultivating the land he faced the problem of unwanted vegetation invading his patch of cereals. Weeds!

Human invaders followed, importantly the Romans who not only built military roads and defence walls against the marauding Picts and sea-rovers but splendid houses around which they created gardens. To make their gardens they brought with them trees, shrubs and flowering plants new to Britain, some of which escaped into the wild, invading the existing vegetation. Among them, as we know from seeds found in Roman remains, were the Creeping Buttercup, the Field Penny Cress, Chickweed, the Bramble and the Field Ox-eye Daisy. Weeds!

It might be thought that an alien cannot compete with plants already well established; but this is not so, for the good reason that a new plant is not attacked by its native pests and predators and consequently there is little destruction of its seedlings. It is not until kind competes with kind for the same food in the same environment that the struggle for existence begins and only the fittest survives.

So it is the very nature of a weed to be aggressive, and when weeds invade your garden they are like these early intruders which, not competing with their own kind in the wild or struggling against their natural enemies, are able to usurp the growing-space of plants that were there before them.

The spread of Christianity brought the monks and their gardens of medicinal and culinary herbs, some of which seeded afar, eventually to become part of the native flora and return to gardens as weeds. We have the monks to thank for one of our worst weeds, Ground Elder, which presumably was grown to cure the gout suffered by the bishops, or why is it also called Gout Weed and Bishop's Weed?

When the first great Elizabeth came to the throne her first service to the nation was to reorganize the navy by building a fleet of new ships, well-armed, that made England Mistress of the Seas; and from voyages to new lands they brought back new weeds among their merchandise.

Security 'against infection and the hand of war' also meant that instead of fortress castles surrounding a piece of land, part of which was always used for growing herbs, undefended gardens could now surround a dwelling. Pleasances were added to the vegetable garden and orchard, and the call went out to the merchants to bring back new plants to decorate them. A list of what was growing in John Gerard's garden in 1599 tells us that 'exotics' had come from as far away as China, but it was not until the 17th century that a professional plant hunter came on the scene, the great John Tradescant who travelled the world to find new plants, succeeded by his son John who brought home rich harvests from England's first colony of Virginia. Among the plants from the New World were, inevitably, some that became new weeds: *Anaphalis margaritacea*, Pearly Everlasting; and the Thorn Apple, *Datura stramonium*, to mention two. Two others he introduced did not, fortunately, survive in their new homes: the Poison Ivy, *Rhus radicans*; and Hemp Dogbane, *Apocynum cannabinum*.

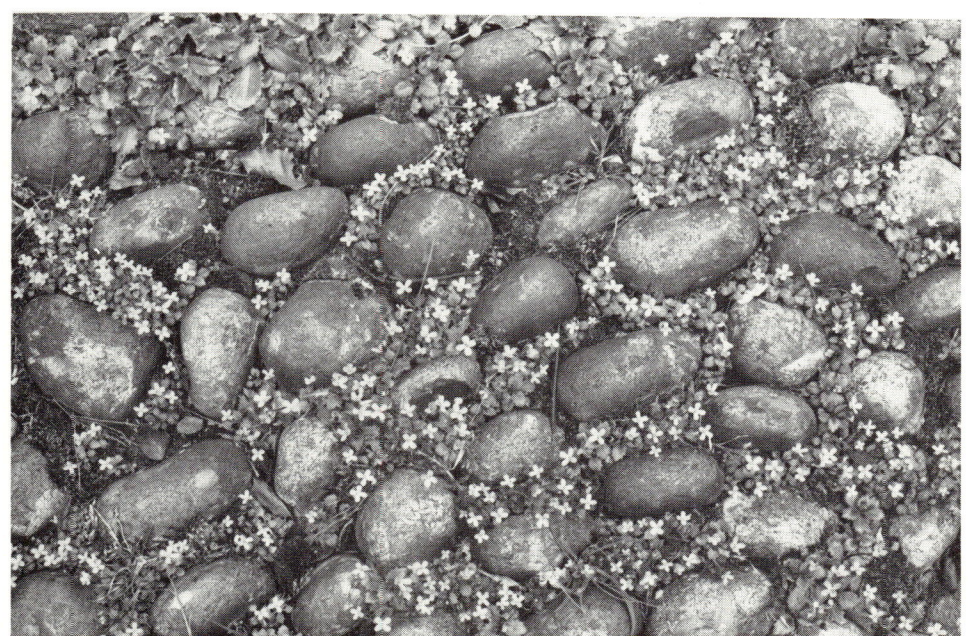

Wherever there's a crack or crevice, weeds will find a way

The curious fact is that few plants introduced from America have become weeds. The explanation is that the plants in the woodlands of Eastern America belonged to relatively primitive families. Having been established for a long time, and adjusted to the same set of conditions, they lacked the ability to meet new and changing conditions such as cleared land. The European weeds on the other hand have come from more modern and aggressive groups. Hence their success in America, the non-success of most American weeds in Britain.

How Weeds Came to America

It is a different story when we come to look at the number of weeds introduced into America from Britain. In his book *New England's Rarities Discovered*, John Josselyn in 1672 gives us a list of those that had become common and yet were unknown in their new country before the landing of the Pilgrim Fathers only fifty years previously. The list includes Couch Grass, Shepherd's Purse, Dandelion, Groundsel, Sow Thistle, Stinging Nettle, Mallows, Wormwood, Chickweed, Mullein, Knot Grass, and Comfrey. Yet another was the Plantain which, Josselyn noted, the Indians called ' "Englishman's Foot", as though it were produced by their treading.' To start life in the new colony the settlers took with them all sorts of herbs, cereals and vegetables, some as seeds, some as living plants in boxes, baskets and pots wherein must also have been the seeds of many weeds.

The New World soon had its nurserymen who imported and distributed European plants. Among the first importations was *Berberis vulgaris*, the common

Bees like Campions: gardeners don't – each plant produces bladders containing about 6,000 seeds

Barberry, as an ornamental and fruiting plant — which became a scourge. It was discovered that wheat growing near the bushes did not thrive, and after a controversy lasting nearly three centuries when stem rust was causing a loss of at least a hundred million bushels of wheat a year, the culprit was discovered in the Barberry. The Wheat Rust fungus needs two hosts for the two separate stages of its development: the Barberry provided the second. The fungus could not breed on wheat alone.

Introductions were sometimes accidental. Earth was used as ballast in ships coming from Europe, and this was put to a new use in the building of transcontinental railroads. The earth contained new weeds which were able in this way to spread over long distances.

By far the most American weeds are of European origin. In a list of 89 weeds common in the State of New York in 1953, just over 40 per cent are common weeds in Britain. In Canada, of 80 of the more important farmland weeds 60 per cent were introductions from Europe and included many of the worst pests. An analysis was done in 1945 for the area of eastern temperate North America. More than half the weeds were of European or Eurasian origin.

Weeds have come to America from other countries too, of course. Hay and straw as packing materials brought weed seeds from many parts of the world. A glance at the list of weeds compiled by the Weed Society of America tells us that the Velvet Leaf (*Abutilon theophrasti*), one of the Mallows, is naturalized from India, that the Giant Foxtail (*Setaria faberi*) came from China in 1931 and probably with the seed of Chinese millet, and that the poisonous Halogeton or Barilla (*Halogeton glomeratus*) was introduced from Siberia about 1930.

So, as man moved, and moved his merchandise, his weeds moved with him.

Garden Plants That Have Become Weeds

It was inevitable that some of the exotics brought to Britain in all good faith by the plant hunters should acclimatize themselves only too well. We think back to the 17th century when John Tradescant the younger visited Virginia and claimed as a prize a beautiful herbaceous plant throwing up dazzling golden spikes of flowers. 'Virga aurea Virgine', he called it. Linnaeus named it *Solidago canadensis*, and we call it Golden Rod; it is certainly now a weed in the United States and although it is still grown as a border plant in Britain, it is one to be halved and quartered lest it ramp too far.

Rosebay Willowherb has become a garden weed. Yet in the late 18th century when William Curtis was publishing *Flora Londinensis*, it was still regarded as a garden escape, previously treasured. From southern Europe in 1648 came the Purple Toadflax, *Linaria purpurea*, which one is pleased to have — at first — for its delicate blue-purple spikes and neat erect habit. Then comes the day when you find it has seeded everywhere. In the dead days of winter when it is hard to find a flower in bloom your heart may be cheered by the sweet scent of Winter Heliotrope, *Petasites fragrans*, and the sight of its heads of lavender flowers braving the cold. Introduced into Britain as a garden plant in 1806 it has become one of our worst tormentors, its persistence far outweighing its charms. It will monopolize any growing-space, overshadowing above-ground, undermining beneath.

I was once foolish enough to accept a young plant of *Chrysanthemum parthenium*, the Feverfew, which made a brave show with its white daisy flowers and fragrant leaves: I know better now. Similarly I once encouraged Snow-in-summer till I found it blanketing some precious alpines and tore it out by the yard. It seemed sacrilege to uproot Violets. Two years later few Primulas had escaped their smothering growth and I had to ban them to a place of their own. The Yellow Fumitory, *Corydalis lutea,* should also have a place of its own where it will reward you with its feathery leaves and flowers in bloom all summer. But it is a terrible seeder and must be contained. Worse is the Grape Hyacinth, *Muscari atlanticum (racemosum)*, whose leaves are bothersome almost all the year round. It will spread like the plague by its small white bulbs. The same can be said of the Nodding Star of Bethlehem, *Ornithogalum nutans*, which is heartbreaking to remove because of its exquisite white-and-green spikes of bells that last so long in water. It is guaranteed to choke the life out of roses, and you will never get rid of it. Dig it out by the spadeful, it will come up next spring as cheerfully as ever. But at least its leaves die down in the summer.

The flower-spikes of lupins should be cut down before the seeding stage, otherwise lupins will quickly become a weed, and this applies to the Tree Lupin, though I always let some of them seed and they always breed true — unlike the herbaceous-border ones which, promiscuous in their habits, can have horrible children.

Ornamental grasses are being grown more and more for their value to the flower-arranger. Again, they should be culled before they seed, otherwise you will find Quaking Grass and Hare's Tail everywhere.

One would hardly think of trees as weeds. Sycamore of winged fruits can easily become so, as can the Ash, and one never seems to detect them until their seedlings are a sturdy foot high. Elder bushes also have a habit of growing up behind your back, so to speak, and the suckering Snowberry, *Symphoricarpus rivularis*, can be a tyrant.

Another shrub that colonizes all too freely is the common *Rhododendron ponticum*. Established bushes spread vegetatively and are costly to get rid of or even keep in check. At Culzean Castle in Scotland where they had become horticultural cannibals they had to be hacked away, sawed, bulldozed and dynamited. Yet what a treasure this rhododendron was thought to be when introduced at Culzean and other unsuspecting gardens in 1763!

You will agree that the pink-flowered *Oxalis corymbosa* with its bulbils, and the yellow-flowered creeping-stemmed *Oxalis corniculata*, are not to be tolerated, especially if the latter gets into a lawn. But what of the Creeping Bellflower, *Campanula rapunculoides*, whose blue helmets are so beautiful? Are you going to heed the warning that its stolons will soon produce a rampant carpet? And will you let the hyacinths seed until they become an embarrassment of riches?

What is one man's flower is another man's weed. In the long run it is up to yourself to decide.

New Weeds To Plague Us

For of course it doesn't stop there. New weeds can arrive tomorrow. As transportation reaches farther and farther, as more and more people travel, so weeds seize their opportunity to hitch-hike their way from continent to continent. Such is the way of a weed. They are equipped for travel, and travel they do. Does one ever discover in a garden a new breathtakingly beautiful flower that has sprung from where you do not know, a plant which, if exhibited, would be given an Award of Merit for its sheer first-time perfection?

But you may find a new weed.

An example of this happened two summers ago when friends in a nearby village summoned me to identify a strange plant that had appeared in their garden. It was not unbeautiful (but did it seed and seed the following year!). At two feet tall it had purple stems and tufts of crimson flowers like those of a larch. It turned out to be *Kochia scoparia*. Why? The garden was in England, the weed not listed as British, while the U.S. Department of Agriculture's weed book declared it to be a serious nuisance in the Plains States and spread throughout most of the northern half of the United States excepting Washington, most of Oregon, and parts of Idaho and Montana. Had my friends been entertaining visitors from a USAAF base? No. Had an American friend been staying? Why, yes! From New England. The distribution map of *Kochia scoparia* included New England. One hitch-hiking seed.

A new weed can easily be introduced along with a smuggled plant. New weeds can seed from rubbish tips where packing materials have been dumped. A good crop of the poisonous Jimson Weed resulted when someone I know shook out a far-travelled suitcase. In America the recently introduced German Velvet-grass, *Holcus mollis*, is spreading like wildfire. Imported and impure seed is providing Britain with some unwelcome guests: from North America has come the parasitic Dodder *Cuscuta campestris* that attacks tomato plants and carrots. Despite the fact that it is a recent introduction and seems not to ripen its seeds in Britain, it is now quite

Helmeted army of invaders: *Campanula rapunculoides*

frequent. It attaches itself by means of suckers and thick orange threads, to feed upon and finally strangle its host.

Bird seed on bird tables is another source of aliens. Seaweed when used as a mulch often contains plant débris such as various grasses. A number of unpleasant weeds are introduced to gardens in container-plants bought from nurseries. Miniature veronicas and cresses move around in this way, as does the almost ineradicable South American *Oxalis corymbosa* which has spread into the United States. The Fiddleneck of orange-yellow flowers, *Amsinckia intermedia*, a relatively new introduction in linseed from Canada, is already well established in Britain and spreading quite rapidly.

That great gardener E.A. Bowles, whose five acres of beauty at Enfield in Middlesex became world-famous, was plant-hunting near Biarritz in 1894 when he came across the spectacle of a pond turned brilliant scarlet by the floating fern *Azolla caroliniana*. 'This wonderful water-fern,' he wrote of it. He sent it home and on his return was delighted to see that it had increased in beauty, its delicate, velvety, finely-cut fronds green and crimson — and fascinatingly interesting to play with, as he remarked, to push under the water and watch arise out of it as dry as ever, but perhaps carrying a few drops on the fronds, to glitter like diamonds in sunlight. In his pond it grew and grew. Before June was over wheelbarrow-loads of it were being removed in order to see the water. When winter came it froze into the ice; and when the ice melted, it looked lovelier than ever, just a more brilliant crimson. It got into the rock garden pools and refused to be evicted, and into the loop of the New River which flowed through his garden. Fortunately next winter's severe frost killed it stone dead. But its cousin *Azolla filiculoides* is now naturalized in the inland waters of southern England.

Farmyard manure, imported topsoil, shoddy, soya-bean waste, sewage works, fodder, freight trains, and even innocent-looking bags of compost: all have been known to bring new invaders. They will continue to come. It behoves us, on finding in our garden a new plant that has all the look of a weed — tunnelling roots, small flowers producing many seeds — to find out about it. Who knows? It may be as harmless as the Orange Balsam, *Impatiens capensis*, that now beautifies miles of the Norfolk Broads. Or it may live to plague us. It is worth reminding ourselves that most weeds are not native to the countries they annoy. For it is the way of a weed to find new fields to conquer, and that metaphorical field can be your garden.

2

Good Weeds and Their Uses

Weeds That Have Become Good Garden Plants

It would be too sweeping a statement to say that *all* weeds can be put to a good use. It depends, of course, on what you call a weed. What is one man's weed can become another's good garden plant. Certainly all the plants we treasure in our borders came originally from the wild, and in their introduced state were sometimes weedy-looking things. What of the original Michaelmas Daisy that John Tradescant the younger introduced into Britain from Virginia in the 17th century? What of his Phlox? We have the hybridizers to thank for converting them into plants that make our autumn borders glorious with rich purples and crimsons and all the shades between.

What of the weedy little Heartsease? In 1687 John Evelyn the diarist, who was a great horticulturist, was growing it in his garden at Sayes Court, near Deptford, east London. But it was not until 1810 that the flower was looked at as more than a little tricoloured wildling. At that date Lady Mary Bennet began collecting Heartsease from various parts of her father's estate at Walton-on-Thames. She made a small garden in the shape of a heart, in which she transplanted all the different kinds she could find. William Richardson, her father's gardener, helped her and began crossing them. His improved varieties came to the notice of James Lee, the famous Hammersmith nurseryman, who imported a blue variety from Holland and began hybridizing them with the Walton Heartsease. Others entered the field. The result was the beautiful modern Pansy of velvety petals.

In California the man who was probably the greatest of all hybridizers, Luther

Collective noun: a choir of archangels

Burbank, looked at three daisies: the Ox-eye Daisy growing on the hills of Massachusetts (*Chrysanthemum leucanthemum*), which was small, tenacious and hardy; an English Michaelmas Daisy which was larger and coarser in stem; and the Japanese daisy (*Chrysanthemum nipponicum*), not large but of exquisite and almost dazzling whiteness. He decided to marry the three and create a queen daisy that would have a slender stem, but firm, at least two feet tall and free from branches, with a flower larger than any daisy ever seen before, and petals of the purest white. He cross-pollinated, selected and re-selected. From 300,000 seeds he grew 100,000 possible candidates, and selected again. The seeding and selecting process was repeated for eight years, and he worked not only for a queen among daisies but a plant that would grow just as well in Alaska as in Florida, Norway and Italy, for all sorts of soils and climates; one, too, that would remain fresh in water for as long as six weeks. At last, after growing millions of daisies, he achieved his queen, a flower of remarkable beauty with brilliant white petals of great size, the centre pure yellow, the stems long and graceful. He named it for Mount Shasta, which he had always revered, the snow-capped peak of the high Sierras which is one of the conspicuous landmarks of California.

Burbank created many new flowers out of weeds. He developed the Amaryllis from a flower only a few inches in breadth to one nearly a foot across and in every shade of crimson, pink and scarlet, that bloomed all the long Californian summer. Working with *Papaver pilosum* and *Papaver somniferum* he produced a fantastic brilliantly scarlet poppy with pointed petals and a purple centre.

Dutch nurserymen have been skilled developers of the hyacinth and tulip, as well as other bulbous plants, for hundreds of years. From the wild *Hyacinthus orientalis* (perfect in a woodland but running riot in a garden) have come the gloriously scented Roman Hyacinths. From the wild species classed together as *Tulipa gesneriana* came the feathered, flamed and parrot varieties for which people beggared themselves to buy a single bulb. The 'tulip fever' reached its peak between 1634 and 1637. Since then, the tall beautiful Darwins and others have been bred for our delight.

In England, Alan Bloom (no fitter name for a flower breeder) has been experimenting for years with weeds and wildflowers. From the invasive *Achillea millefolium* he has raised his 'Cerise Queen'. He has given a new dignity to the Spotted Dead-nettle and removed the wild Mallow from the category of weed by developing 'Primley Blue'.

Sometimes Nature herself produces a freak among her wild plants, as she did with a double buttercup and a variegated-leaved Water Figwort. Both were adopted as garden plants.

Nobody yet, to my knowledge, has thought of developing that weed the Dandelion which, robbed of its parachute power, improved in size and given a scent, might make an attractive garden plant as well as a good sub-soiler. Luther Burbank induced a sweet-smelling scent into the malodorous Dahlia, so why not into the Dandelion? Nor has the Grass of Parnassus been considered, with its upturned chalice of green-veined ivory petals. Nor the enchanting vetches, nor the Stitchwort of starry white flowers — a mere wayside weed, but a potential garden treasure.

Why not? Perhaps their day will come.

Good Companions

There are some weeds that make good companions for other plants, whether in flower borders or in the kitchen garden. Shakespeare knew this. In *Henry V* he wrote:

> The strawberry grows underneath the nettle,
> And wholesome berries thrive and ripen best
> Neighboured by fruit of baser quality.

The subject is currently arousing great interest. Though not yet critically tested, something is already known about it. It has been found that roses benefit if alliums are grown near them: black spot is reduced. In some countries where roses are grown commercially for making perfume, Garlic (*Allium sativum*) is grown with them. We can choose ornamental species, such as *Allium moly* and enjoy its large brilliant yellow blooms, or *Allium schoenoprasum* var. *sibiricum* for its rose-purple flower-heads. The Foxglove makes another good companion, encouraging neighbouring plants and stimulating their growth and hardiness. The Wild Chamomile (*Matricaria chamomilla*) was called the 'plants' physician': a sickly plant grown near it soon recovers.

There is a scientific explanation as to why some plants, including weeds, make good companions, while others are enemies of their neighbours. The various compounds that plants excrete through their roots have a direct influence on the soil, some stimulating the growth of valuable organisms, while others poison the beneficial bacteria. The effect of their exhalations is another factor: their scents of volatile substances which make them welcome or repellent to other plants.

Roots are important, for apart from what they do by excreting useful compounds they have another use. Nature sees to it that good companionship is maintained below-ground as well as above, by means of a diversity of root- and stem-systems, of which there are four kinds: tap-roots; fibrous roots; creeping rhizomes; and bulbs, corms and tubers. The soil of a neglected garden will be compacted, in rainy weather collecting pools of water that never soak down to the thirsty sub-soil. What does Nature do? She sends in seeds of deep-rooted weeds, the ploughmen of the plant world: thistles and umbellifers like Cow or Wild Parsley which aerate the soil and bring to the surface valuable plant-food materials. Plantains, docks and buttercups will follow, then the shallow-rooted chickweed and Groundsel. When Groundsel appears, the soil is in good heart again.

This is the way to plan your garden, taking a leaf out of Nature's book by mixing plants with different root-systems. In a dry sandy soil shallow-rooted annuals will not do well. They will do better if companioned by deep-rooted perennials which by their capillary action bring up the moisture. Taking the tap-rooted Cow Parsley as your model, use lupins in a poor soil: they will not only bring up moisture but add calcium as well.

Weeds, it is known, generate warmth, and spurges preserve the soil's warmth and so help and protect plants that are tender, which come from a warm climate. We can let the Petty Spurge or Sun Spurge go unweeded near them, or we can substitute attractive garden spurges.

And what of the kitchen garden? Though a weedy vegetable patch can be an untidy sight, certain weeds should be tolerated for the good they do. Don't uproot dead-nettles growing near your rows of potatoes: they benefit them. Yarrow helps most vegetables. F.C. King, for many years in charge of the famous garden at Levens Hall in England's Lake District, found that the best way to secure a good crop of sound onions was to allow weeds to develop in the onion bed after about the first week in July. The growing weeds, by denying the onions a supply of nitrogen, improved their keeping qualities, and by digging in the weeds in the autumn provided a supply of humus for the next crop. 'Dr Chamomile' is good to grow with cabbages, savoys and all the other brassicas.

So in these ways both above-ground and below-ground Nature achieves and maintains fertility. When we imitate her we achieve the same good results. The process is known as bio-dynamic gardening.

Bio-chemists are still learning the alphabet of plant associations, and one day they will give us full and factual information about them. Meanwhile gardeners can learn much by observing how plants are affected by what weeds spring up near them. And if they cannot tolerate these weeds, we shall see there is another way to use them, by which the good that is in them can be returned to enrich the soil.

Spurges warm the earth, but this one – the Petty Spurge – is a weed. Grow ornamental Spurges instead.

The Compost Heap

This brings us to that most important part of the garden. It could be likened to your kitchen, where good meals are prepared and cooked to feed your family. The compost heap does just that: it feeds the soil of your garden with all the nutrients plants need for growth, health, and a good constitution.

Our planet's soil took millions of years to form, during which time plants were living and decaying, returning to the soil what they had taken up, a cycle of life by which fertility was built up and maintained. It follows that if the gardener does not adopt this very important law of return, the soil will become depleted. Dust bowls are created in this way.

It is mainly in the top six or nine inches of soil that plant-roots grow, and in that top layer is a wonderful world of workers living in partnership with the plant-life above. Worms are tunnelling in search of food: they consume quantities of fallen leaves, dragging them down to their burrows and tearing them into the finest shreds. Finally the leaves are turned into a rich humus by having passed through their alimentary canals, and this humus is brought to the surface in the form of casts. Worm-burrows provide drainage and allow air into the ground, both seeping indirectly through to the soil. The earth-loosening worms also help the roots of plants to travel down more easily. Charles Darwin wrote a book in praise of earthworms. He had this to say of their ploughing activities: 'It may be doubted whether there are many other animals which have played so important a part in the history of the world, as have these lowly organized creatures', pointing out that the whole of this planet's covering of soil passes every few years through their bodies.

There are also bacteria in the soil that convert nitrogen into a form plant-roots can use, without which most plants would suffer from nitrogen starvation. Other bacteria decompose organic material. Together these creatures make up a vast population busied in the work of soil-building.

But all this takes time. The good gardener speeds up the process by making a compost heap. This, when properly made, generates enough heat to decompose the material quickly.

There is no better way to soil fertility than an abundance and variety of weeds in the compost heap. Besides weeds, virtually everything that has once lived can be used: lawn mowings, leaves, thin prunings; kitchen waste such as coffee-grounds, tea-leaves, vegetable-peelings and vacuum-cleaner dust. Cabbage stalks and Brussels sprout stumps should first be smashed. If yours is an acid soil, crushed eggshells can be added. NOTS are evergreens like laurel leaves and ivy which will not compost easily and should not be used. Neither should pine needles, rhubarb leaves nor moss.

To contain the compost, first make a box, with sides of 3–4ft (90–120cm) and up to 5ft (150cm) high, using anything readily available such as breeze blocks, bricks, baled straw, timber, corrugated iron sheeting. Then, lightly fork over the ground. This provides drainage and helps earthworms move into the compost at the appropriate time.

Having made the box —
1 Lay a foundation of brushwood and land drainage pipes or double brick rows, for bottom aeration. Good bottom aeration is MOST IMPORTANT. For

when the cold air enters, the heat is drawn up like a flue, warming the whole heap. Side ventilation is not necessary, and nowadays is not advocated.

2 Thoroughly mix the weeds and waste materials, watering them if dry but not making them sodden.

3 Spread a layer of 6–9in (15–22cm) of the mixture, evenly scattering an activator on top. The following are all good: fresh-cut nettles, animal manures, non-toxic sludge from non-industrial areas, dried blood, bone meal, hoof and horn meal, powdered seaweed, liquid seaweed extract greatly diluted, or a few sprinkled handfuls of sulphate of ammonia. Garden centres sell proprietary inorganic and organic activators. All these provide nitrogen for the soil micro-organisms to work on. Then, just a sprinkling of top-soil is necessary. Bonfire ash can be substituted for top-soil. Either helps to hold any liberated ammonia till the micro-organisms can make use of it.

4 Build up the next 6–9in layer of pre-mixed material, as before finishing with a sprinkling of top-soil. In town gardens the soil is often acidic and the addition of lime is helpful, scattered at the rate of 4oz to the square yard.

5 On top of the next layer add another scattering of activator and top-soil, and continue with these layers until the box is full. As you build up, so as to retain the heat that will start generating, spread on top a protective layer of sacks or an old carpet, or a sheet of polythene punctured with a few holes. These holes are important: they release the carbon dioxide. Finally, level off and spread a layer of top-soil. Then, against rain, place a sloping lid on the four corner-posts.

A temperature of 140°F (60°C) should be aimed at for fast breakdown and for destroying seeds and any diseased material. To test the heat, thrust a bamboo cane down into the heap, withdraw it and have ready a thermometer (a cooking one, not

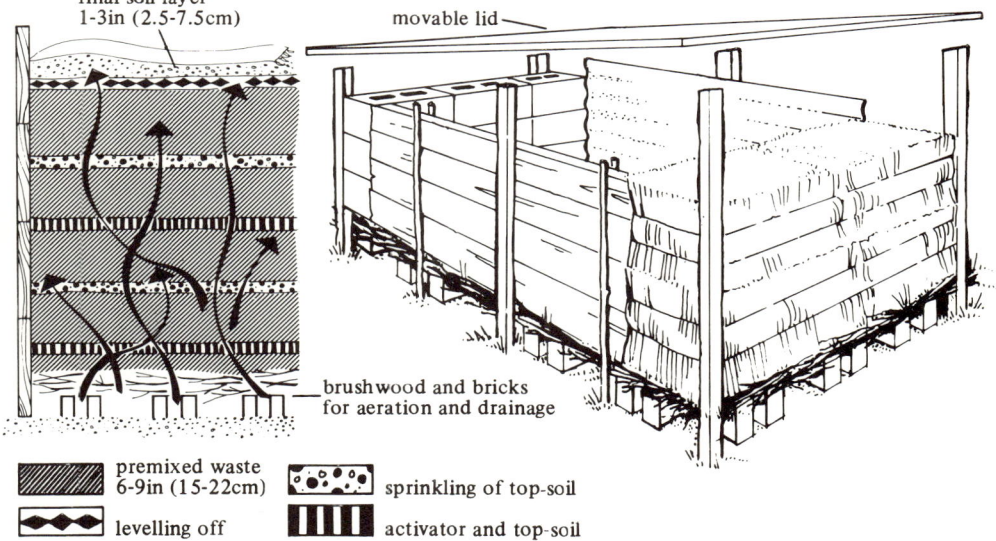

final soil layer
1–3in (2.5–7.5cm)   movable lid

brushwood and bricks
for aeration and drainage

premixed waste
6–9in (15–22cm)   sprinkling of top-soil

levelling off   activator and top-soil

Pattern for Queen Anne's Lace

a clinical) to which a string has been tied. Lower this down into the hole and let it remain for a few minutes. A grain thermometer can also be used.

Within a week of completing the compost heap the centre should have warmed up to its maximum temperature. On lifting the blanket, steam should rise from the top which will feel definitely warm. The heat extends through the heap to about 6in (15cm) from the sides, and in this outer 6in the material composts slowly, so that seeds and any diseased material may not be completely destroyed. The heap should therefore be turned 3–4 weeks after completing it, bringing the cooler top, bottom, and sides to the middle. But if this cannot be done, the outer 3–6in (7–15cm) should be sliced off the compost before use, and this partly decomposed material used to start the heap anew.

In summer the compost should be ready in about 12 weeks. Winter composting takes longer: a heap made in October–November will not be ready till March–April, though in southern climates it will be faster. Mixing is important, especially with grass cuttings. These should never be dumped without mixing or they will become a slimy black mass and stop the process of decomposition.

Good compost smells good. It is friable and if squeezed in your hand will tend to hold together. The best time to apply it is in spring, so that the nutrients are readily available for the plant roots to pick up. But if a soil-conditioner is needed, use it as an autumn mulch.

Composting borders means you can stop digging them. Spread it on the surface and lightly fork or rake it into the top few inches. Leave the rest to the worms who will soon work it in. A bucketful to a square yard is a good average dressing, but more will not hurt.

You have now put back all the nutrients the weeds have collected, to produce strong and healthy garden plants.

'Good Companion' note: plant elder bushes nearby: they help fermentation. Do not put the compost heap under conifers: they impede the process.

Weeds As Soil Chemists

It is important to know what kind of soil your garden has — first, whether it is acid or alkaline. You will then know what kinds of plants you can grow successfully. If, for instance, you long for great banks of gorgeously-flowered rhododendrons, you must banish the thought if yours is an alkaline soil: you will have to put up with gorgeously-flowered *Cistus* instead. Weeds are good indicators. The presence of some of the following tells that your soil is acid:

Spurrey
Corn Chamomile
Henbit
Sheep's Sorrel
Wild Radish
(American names in brackets)

Black Bindweed
Small Nettle (Burning Nettle)
Annual Mercury (Boys-and-Girls)
Common Stork's Bill (Alfileria)
Shepherd's Cress

These indicate an alkaline soil:

Common Mouse-ear Chickweed
Field Mouse-ear Chickweed
Fumitory
Corn Poppy

Wild Carrot
Hoary Plantain
Night-flowering Campion

Some weeds, however, will take to any soil; and some of the same genus will prefer one to another, the Scarlet Pimpernel thriving on lime-deficient acid soil, while the blue-flowered variant *Anagallis foemina (caerulea)* likes a lime-rich soil. Litmus paper will give a surer indication of an acid soil, staining the paper red if it is pressed on to the soil when moist. The paper must not be pressed down with the fingers, even if they are clean, for fingers are quite acid.

The best way to test for acidity and alkalinity is to take soil samples from different parts of your garden, label them, and despatch them to your local horticultural adviser. He will send you an analysis based on the pH scale (pH means hydrogen ion potential), being the balance between hydrogen ions and hydroxyl ions.

Neutrality, neither very alkaline nor very acid, is pH7. Below pH 6.5 the soil is acid, and pH6 is the point where only acid-loving plants such as rhododendrons and

most heathers will thrive. If the figure is up to pH8, the alkalinity is such that
certain essential foods are locked up and plants will show signs of starvation. The
scale is logarithmic, so that pH5 is 10 times as acid as pH6, and pH4 is 100 times as
acid as pH6.

Fortunately, apart from the definite lime-haters, most plants will thrive in
neutral or slightly acid or slightly alkaline (chalky-limy) soil. These are indicators
of a good, balanced loamy soil:

Common Forget-me-not
Coltsfoot
Field Milk-thistle (Field Sow-thistle)
Spiny Milk-thistle (Spiny Sow-thistle)
Stinking Mayweed (Mayweed)
Curled Dock (Yellow Dock)
Creeping Thistle (Canada Thistle)
Goosegrass
Yarrow
Field Penny Cress
Sun Spurge (Wartweed)
Long-headed Poppy
(American names in brackets)

If plenty of weeds grow in your garden, take heart — no soil of any worth is free
of them: it proves your soil is good. And the more kinds of weeds that grow in it the
better you can be pleased. For a good soil has many different elements. The big three
are the food-plant fertilizers nitrogen, phosphorus and potassium, with smaller
amounts of calcium predominating over the trace elements of copper, iron,
magnesium, cobalt, silica, manganese, sulphur and boron, with other ingredients
depending on the underlying rock and mineral content. Garden plants need all these
to make them grow and keep them healthy. Weeds absorb them greedily. But put
the weeds in the compost heap and the nutrients they have mined from deep down
or have filched from other plants will go back into the soil.

The chart opposite shows which weeds are particularly rich in these nutrients and
trace elements.

The Daily Diet of a Weed

NITROGEN

All-seed
Bindweed
Black Nightshade
Broad-leaved Dock
Chickweed
Clovers
Common Storksbill
Creeping Thistle
Dandelion
Fat Hen
Groundsel
Knotgrass
Procumbent Yellow Sorrel
Purslane
Redroot Pigweed
Redshank
Sow Thistle
Stinging Nettle
Vetches
White Campion
Yarrow

PHOSPHORUS

Broad-leaved Dock
Bulbous Buttercup
Fat Hen
Purslane
Tufted Vetch
Yarrow

POTASSIUM

Broad-leaved Dock
Bulbous Buttercup
Chickweed
Coltsfoot
Corn Chamomile
Couch Grass
Fat Hen
Great Plantain
Purslane
Stinging Nettle
Tansy
Thistles
Tufted Vetch
Yarrow

CALCIUM

Coltsfoot
Corn Chamomile
Creeping Thistle
Daisy
Dandelion
Fat Hen
Goosegrass
Great Plantain
Horsetail
Purslane
Scarlet Pimpernel
Shepherd's Purse
Silverweed
Stinging Nettle

BORON

Spurges

COBALT

Bulbous Buttercup
Horsetail
Ribwort Plantain
Rosebay Willow Herb
Tufted Vetch

COPPER

Bulbous Buttercup
Chickweed
Coltsfoot
Creeping Thistle
Dandelion
Great Plantain
Ribwort Plantain
Sow Thistle
Spear Thistle
Stinging Nettle
Tufted Vetch
Yarrow

IRON

Bulbous Buttercup
Chickweed
Chicory
Coltsfoot
Crosswort
Creeping Thistle
Dandelion
Fat Hen
Great Plantain
Ground Ivy
Groundsel
Horsetail
Silverweed
Stinging Nettle

MAGNESIUM

Chicory
Coltsfoot
Daisy
Horsetail
Ribwort Plantain
Silverweed
Yarrow

MANGANESE

Bulbous Buttercup
Chickweed

SILICA

Couch Grass
Great Plantain
Horsetail
Knotgrass
Stinging Nettle

SULPHUR

Coltsfoot
Fat Hen
Garlic
Purslane

3

Bad Weeds –
Control and Destruction

These Are For Burning

Practically all weeds can be composted, even suspect things like Couch Grass with its creeping underground stems — IF the compost generates sufficient heat ($140°F$ ($60°C$)). But, as with all generalizations, there are the exceptions. These should be burned:

Weeds with bulbils (e.g. Lesser Celandine, Corymbed and Broad-leaved Sorrels)
Weeds obviously diseased
Weeds treated with chemical weed-killers or pesticides
Moss
Ivy
Bramble
Fungi
Mowings from grass treated with weed-killers
Grass sods
Unsmashed stalks of thistles, cabbage, Brussels sprouts

IF your compost heap does not cook well, burn the following because they do not rot down easily or quickly:

Weeds at seeding stage
Underground parts of rhizomatous and stoloniferous weeds (e.g. Couch Grass, Ground Elder or Goutweed, bindweeds, Horsetail, Creeping Buttercup)

Tubers and tap-roots (e.g. of Cow or Wild Parsley, Cow Parsnip, Hemlock, Velvet Leaf)

Some weeds act as hosts for plant pests. The following are notorious:

Annuals
Fat Hen (Common Lamb's Quarters)
Yorkshire Fog (Velvet Grass)
Charlock
Annual Meadow Grass (Annual Blue Grass)
Wild Radish

Overwintering
Groundsel
Penny Cress
Ribwort
Shepherd's Purse
Chickweed
Black Nightshade (Common Nightshade)
Great Plantain (Broad-leaf Plantain)
Knotgrass
(American names in brackets)

All these should be burned. But as you will have seen from the chart on page 33, some of the above weeds contain useful nutrients and minerals. So do you compost these for the good they can do? Or destroy them as possible pest-carriers? Ones that overwinter are more likely to harbour pests which seek transit hotels at this time of year. The stray Groundsel and mat of chickweed should therefore not be left in the ground from autumn onwards. The ones that can harbour the Club Root fungus (Charlock and Wild Radish) should be burned immediately. Further details are below.

If you are in any doubt, it is safer to burn than be sorry. Bonfire ash makes an excellent addition to the soil (except near mint). A scattering round a rose bush acts as a beauty treatment. It is good too as an additive to the compost heap. So all is not lost that is burnt.

Weeds As Hosts for Plant Pests

There are many reasons why we should keep our gardens weed-free: not only because we need the space they occupy for growing border plants, and because they filch from them the nutrients and minerals these require; nor just because they 'spoil the look of the place', but, very importantly, because they can harbour pests and plant diseases, especially during the autumn and winter.

There are the harmful insects. Among them are the aphids, classed by the entomologists as 'true bugs'. In a weed survey carried out in England from 1963 to 1972, 40 per cent of weed samples examined during April were infested with aphids. These feed on weeds before moving on to attack herbaceous plants and vegetables.

Do not harbour Groundsel – Groundsel harbours pests

The Black Bean Aphid (*Aphis fabae*), for instance, commonly known as Blackfly or 'The Blight', can overwinter on Fat Hen and Black Nightshade, migrating in the summer to beans, beetroot and spinach. *Myzus persicae*, the Peach-potato Aphid, one of the Greenflies (there are 500 different species of Greenfly!), overwinters on chickweeds, Groundsel and Fat Hen, and attacks lettuces and other crops. Its cousin *Myzus ascalonicus* lives on plantains and attacks strawberries and shallots. Mature aphids suck the cell-sap of plants and thus check their growth; they cause leaves to become rolled or blotched, or galls to develop. Recently they have been proved to cause the spread of virus diseases from one plant to another.

Virus diseases include the various Mosaics like the Lettuce Mosaic and the Cabbage Black Ringspot. These are harboured by chickweeds and Groundsel, Fat Hen and Shepherd's Purse.

There are also nematodes and fungi. Nematodes are eelworms, and these attack such things as narcissus bulbs, buds, leaves and stems. They are called secondary invaders because they enter the tissues of plants damaged by other organisms, thereby extending the initial injury and causing the final destruction of the damaged plant. You are unlikely to see them, for they are minute and often microscopic creatures. Some inhabit the soil around roots, especially grass-roots, and they

The aerial roots of the Ivy burrow into the mortar of walls, eventually bringing them down with its weight

overwinter in the roots, even though the weed host has been killed by frost. Others invade dried and dispersed seeds. Again, chickweeds, Groundsel, and Fat Hen are the commonest hosts, with Shepherd's Purse, Knotgrass and Annual Meadow Grass. Between them these weeds are responsible for nurturing nematodes that damage root vegetables, onions, lettuce and brassicas, strawberries and blackcurrants; and in the flower garden dahlias, delphiniums, phlox, roses, rudbeckias and other herbaceous plants.

Finally there are the noxious fungi such as the Wilt Fungus, Powdery Mildew and Downy Mildew, Petal Blight and various Rusts. Groundsel is a host of all of them. Charlock and other Crucifers, Wild Radish and Yorkshire Fog are hosts of the Club Root disease that attacks brassicas, root vegetables and such garden flowers as wallflowers and stocks.

The subject is vast, the number of pests and diseases enormous. Indeed, it has not been fully explored, particularly with regard to which garden weeds act as hosts to what diseases and pests. Here we can only warn the gardener that these things exist and that overwintering weeds are likeliest to threaten his next summer's flowers and vegetables. They should therefore not be dug in but can safely be composted, or consigned to the cleansing flames.

Poisonous Weeds

He that bites on every weed must needs
light on poison.

John Ray's *Compleat Collection of English Proverbs* 1742

Some weeds are poisonous to humans in one degree or another. They may be merely skin irritants: others if eaten may be deadly to children and even adults. The following list, compiled from the weeds described in this book, explains which parts are poisonous and those (marked F) known to have produced fatal results, usually applying if the particular weeds are eaten in moderate amounts.

Weed	*Poisonous Part*
Bryony, Black, White	Root and berries. F
Buttercups	Sap
Canadian Fleabane (Horseweed)	Leaves
Fool's Parsley	All parts. F
Hemlock	All, especially young leaves and unripe fruits. F
Ivy	Leaves and berries
Ivy, Poison	All parts
Jimson Weed	Leaves, unripe capsules, especially seeds. F
Lords and Ladies	All parts, particularly berries
Nightshade, Black	All parts
Nightshade, Deadly	All parts. F
Opium Poppy	All parts, particularly unripe capsules
Poisonous Lettuce	Leaves and stem
Spurges	All parts
Spurge, Petty	All parts. F
(American name in brackets)	

Harmful to animals, particularly to grazing stock, when in the green state, are the following:

Weed	*Poisonous Part*
Dutch or White Clover	Possibly all parts
Field Poppy	All parts
Groundsel	All parts
Docks	Leaves
Horsetail	Possibly all parts
Charlock or Wild Mustard	Seeds
Wild Radish	Seeds
Ragwort	All parts
Sorrels	Leaves
Common or Narrow-leaved Vetch	Possibly all parts

Tendrils of White Bryony

Biological and Chemical Control

We have discussed weeds that make good companions for garden plants. This symbiosis or 'living together' has its reverse in antibiosis: it has been found that some garden plants kill some weeds.

Tomatoes, for instance, have such an antipathy for Couch or Quack Grass that they can be used to kill it. Secretions from the tomato plant's roots have this deadly effect.

In another way nasturtiums kill Horsetail. The competition for light is the important factor here: Horsetail cannot exist if overshadowed, and with its large leaves the nasturtium does just that.

Weeds also annihilate weeds by internecine warfare. The presence of dead roots of Brome Grass inhibits the growth of young Brome Grass plants. Similarly seedlings of the Canadian Fleabane or Horseweed are stunted by the dead roots of the same species. In this case the lethal weapon is a by-product of decomposition.

But by far the most dramatic killer is the Mexican *Tagetes minuta* which despite Linnaeus's unfortunate name for it is the tallest marigold in the world (its maximum height is 12ft (4m)), but Linnaeus was referring to the smallness of its flowers. Experiments by a Dutch nurseryman proved that it killed his Narcissus Eelworm. In 1960 an English gardener discovered that Ground Ivy, Ground Elder and Convolvulus vanished when he grew the giant marigold among them. Couch Grass also disappeared. Since then, other gardeners have reported success with plantains, Shepherd's Purse, Groundsel, Ragwort, Fool's Parsley, docks and Creeping

'Ware the Deadly Nightshade, poisonous in all its parts – leaves as well as luscious-looking berries

Buttercup. Horsetail and nettles have been added to the list, with thistles, Fat Hen and Black Nightshade.

While much scientific work has been done on antipathetic symbiosis, much more remains to be done: we do not yet have a dictionary telling us how we can keep our gardens weed-free by planting specific killers. Nor can we grow the giant marigold in rock gardens where its shade and roots would stifle tiny alpines. Until that day we either hand-weed or use a herbicide.

Using the right herbicide for the job is important. Grasses respond to one group of chemicals, broad-leaved weeds to another. Most of the latter, but not many grasses, are sensitive to 2,4–D, and 2,4,5–T, silvex (fenoprop), dicamba and related compounds. Grasses are primarily destroyed by dalapon, amitrole (aminotriazole) and tetrazine. These names, by the way, do not refer to branded products but to the active ingredients to be looked for on the labels of proprietary weed-killers.

Different herbicides work differently. There are selective and non-selective kinds: the first selecting only those weeds to be killed, leaving other plants unaffected: these are used on lawns, for instance. They act on roots, or on leaves, or on both roots and leaves. The non-selective ones kill everything they touch, these for places like paths and drives. The first three groups below are selective, the last non-selective.

Absorption by roots	dichlobenil, DCPA*, bensulide*, propachlor, simazine, atrazine
Killing by contact with leaves	2,4–D, aminotriazole, dalapon, dichlorprop, fenoprop, silvex, ioxynil, MCPA, dinitro, mecoprop, morfamquat
Both leaf and root uptake	dicamba, dalapon
Killing whole of weed on contact	common salt, sodium chlorate, borates, paraquat, diquat, 2,4,5–T

*Obtainable only in the USA

The first step in using herbicides is to identify the weeds to be killed. You can then choose the best way to exterminate them: that is, the part most necessary to be killed — the creeping rhizomes of Couch Grass, for example, fragments of which, if the plant is pulled up, are left behind and go on producing new shoots; and, if trees are in the vicinity (remember their roots are far-spreading and can absorb poisons), use a foliage-contact weed-killer not a root-killer.

The chart on pages 44–5 tells you which ingredients each herbicide should contain for each purpose. There are important DO's and DON'Ts:

DO keep weed-killers away from the reach of children, preferably in a locked cupboard in the garden shed.
DO follow the manufacturer's instructions carefully.
DO keep a special watering can, sprayer and measure.
DO wash hands and face after spraying.
DO dispose of containers safely once they are empty.
DON'T spray on a windy day or in hot sunshine.
DON'T spray near a fish tank, stream or pond.

Control in Ponds and Lawns

A piece of water, however small, adds interest to any garden. It gives you the bonus of reflections, to say nothing of the charm of the moisture-loving plants you can grow round the margins. Nor will you leave the water untenanted by plants: fish need them for food and shelter. But alas, these pond plants, always referred to as pond *weeds*, tend to behave like weeds, reproducing themselves vigorously until not an inch of water is visible.

There are preparations on the market for controlling water plants but amateur gardeners are not recommended to use them. Culling by hand is the best procedure,

The Black Medick, charming growing at a deserted station, a pest in lawns and paving

but if this is beyond you consult your local horticultural adviser, Extension Service or botanic garden. Professionals know what to use and what precautions to take.

The worst offenders are the floating species: Duckweed (*Lemna minor*), the Ivy-leaved Duckweed (*Lemna trisulca*), and Frogbit (*Hydrocharis morsus-ranae*); and of the submerged species the Water Milfoils (*Myriophyllum* species), Canadian Pondweed (*Elodea canadensis*), the Water Crowfoots (*Ranunculus* species). There is also Blanket Weed, which is caused by various algae. This should be removed as soon as possible, as fish can become trapped in it.

Lawns are simple to deal with by the use of selective herbicides for new lawns, others for established lawns, ones particularly for moss, for speedwells, for clovers, and so on. Timing is important: a new lawn should be treated when the first flush of seedling weeds has appeared, usually 4 weeks after sowing, and spraying should be done when the soil is moist.

The weeds most likely to trouble you are Yarrow, chickweeds, clovers, Crab Grass, speedwells, Daisies, Dandelions, Knotweed, Nimble Will, plantains, Sheep's Sorrel, Creeping Buttercup, Silverweed and Black Medick. The picture of this last little nuisance, opposite, shows it growing on a platform of a redundant railway station. It can be equally at home in your lawn!

Before buying your lawn herbicide, again identify the weeds you want to kill. Then you will be sure of buying the right preparation.

A blind man went to buy a field. His horse was led by the would-be vendor, and when they got there the blind man said: 'Tether my horse to the nearest thistle.' Said the vendor: 'There are no thistles in the field.' 'Then tether my horse to the nearest bunch of nettles,' said the blind man, and when this was done the blind man nodded. 'I will buy your field.'

Table of Weedkillers

WEED SITUATIONS	Aminotriazole	2,4-D	Dalapon	Dichlobenil	Dichlorprop Dicamba	Fenoprop Silvex
Established Lawns		●1			●	●
New Lawns						
Rose Beds				●		
Flower Beds Other Than Roses						
Shrub Borders				●5		
Vegetables						
Paths and Drives	●8	●				
For Clearing a Derelict Garden 10	●		●			
Fruit Trees				●		
Cane and Bush Fruits				●		

1 Most weedkillers for use on established lawns are based on either 2,4-D or MCPA to which has been added either dichlorprop, fenoprop or mecoprop to increase the range of susceptible weeds.
2 A mixture of ioxynil and mecoprop is available for the specific control of speedwell in lawns.
3 Paraquat/diquat, in any of its uses, must *only* be applied to the weeds: drift on to other plants can cause severe damage.
4 Propachlor should be used instead of dichlobenil or simazine when the roses are underplanted with bulbs, annuals etc.
5 Consult labels before use as some shrubs are sensitive to these chemicals.

Ioxynil	MCPA	Mecoprop	Morfamquat	Paraquat/Diquat	Propachlor	Simazine	Dinitro	Sodium Chlorate	Sodium Borate	2,4,5-T
	●	●								
● 2			●							
				● 3	● 4	●				
				●	●					
				●		● 5				
				●	● 6	● 7				
	●			●		●		● 9		
						●	●			●
				●		● 11				
				●		●				

6 Propachlor is safe to use on brassicas, onions and leeks.
7 Simazine is safe for asparagus, broad beans and sweetcorn.
8 Aminotriazole is only available as a mixture with simazine and MCPA or simazine alone.
9 Sodium Chlorate may possibly be washed on to adjacent areas if heavy rain falls soon after application.
10 Consult labels. The appropriate chemical very much depends on the growth to be killed. For example, dalapon is excellent against grasses and 2,4,5-T against bramble, elder etc.
11 Not beneath plums or cherries.

4

Know Your Weeds

Spikes!

How To Use the Identification Chart

We come now to the actual identification of the weeds you may find in your garden. Over 200 of the commonest are described. Similar ones are grouped together (Buttercups, Chickweeds, Docks), and each group has a master-drawing, each different weed having a drawing of a detail showing the distinction between it and the other members of the group.

You have a particular weed you want to identify. To find the page of the book where it is described and pictured, you will have to know three things about it in order to match it with the page-finding chart overleaf.

These concern:

1. The FLOWER, the *number of its petals* and whether it grows singly (*solitary*) or in *clusters*. The flower may have an unusual shape, the lower petals forming a *lip*, or it may be in the form of a *trumpet*, or have a *pea-flower*. Other weeds have flowers so small that it is difficult to count the petals or see the flower's shape, and these usually grow in a *spike*. Also in this category are the flowers growing up the stalk but not arranged in clusters. Grasses are separately listed.

FLOWER				
0–3 petals	4 petals	5 petals	6 or more petals	Composite
Lipped	Pea	Spikes	Trumpet	Grasses

2. The LEAF-SHAPE. These shapes fall roughly into 4 categories: 3 dealing with what are called simple leaves, such as heart-shaped or oval; the fourth (called compound) where the leaf is deeply indented into *lobes* or where the leaf is *pinnately divided* (pinnate means 'featherlike'), having leaflets arranged on each side of a common stalk. If the leaf is twice divided, each leaflet having a stalk with leaflets of its own, this is termed *bipinnate*. Some leaves are even tripinnate.

3. The LEAF-ARRANGEMENT, whether the leaves grow up the stem *alternately*; or in pairs *opposite* each other; or in a *rosette* close to the ground; or grow directly from the root; or grow in *whorls* up the stem.

With these three main facts about your weed — which the symbols below will help you select — turn to the page-finding chart overleaf where the boxes refer to page numbers. Here you will be able to match your flower, leaf-shape, and leaf-arrangement with the relevant columns. Several weeds may answer to the same data and you will have to look up several entries. A starred number denotes that the weed has leaves of two different shapes or arrangements.

LEAF SHAPE			
Linear to Lanceolate	Oval to Round	Arrow to Heart	Lobed to Compound

LEAF ARRANGEMENT			
Alternate	Opposite	Rosette or Ground	Whorled

Weed Identification Chart

FLOWER		LEAF SHAPE		
Number of Petals	Arrangement	Linear to Lanceolate	Oval to Round	Arrow to Heart
0–3	Clustered		137, 162, 163	
4	Solitary	136	125, 146, 156, 157, 158	
4	Clustered	52, 53, 78, 79*, 127*, 128*, 173, 174, 175	52, 53, 79*, 175	129
5	Solitary	63, 64, 67	64, 65, 131, 147, 151	
5	Clustered	66, 87*, 109, 115, 165	66, 87*, 130, 132, 147	110, 122
6 or more		90		62
Composite	Solitary	68	81	74
Composite	Clustered	86, 113*, 133*	89, 114	176
Lipped	Clustered		110, 117, 118, 119	116, 117, 119
Pea	Clustered			
Spikes		83, 93, 94, 109, 137, 140, 142, 143, 160, 164, 174	84, 85, 94, 138, 139, 141, 159, 160, 161, 164, 165, 166	58, 60, 84, 94, 120
Trumpet		54*, 144	112, 131	54*, 56, 57, 58
Grasses	96 - 103			

Lobed to Compound	LEAF ARRANGEMENT			
	Alternate	Opposite	Rosette or Ground	Whorled
	137, 162	162, 163		
145, 146, 158	125, 145, 146, 156, 157, 158	136*, 157	136*	
76, 77, 79*, 127*, 128*, 129	76*, 77*, 78, 79*, 127, 128*, 129*, 174, 175	175	76*, 77*, 79*, 128*, 129*, 173	52, 53
61, 62, 70, 71, 91, 92, 121, 153, 154	61, 62, 121, 131, 154	63, 64, 65, 67, 91, 92, 147, 151	70, 71, 153	64
59, 60, 75, 91, 92, 104, 107, 121, 134, 135, 154, 155	59, 60, 87, 91, 92, 110, 121, 122, 130, 132, 134, 135, 154	66, 75, 104, 107, 109, 115, 147	87, 155	165
	90		62	
69, 82, 123, 124, 167	68, 69, 123, 124, 167*		74, 81, 82, 167*	
55, 69, 70, 105, 106, 113*, 126, 133*, 149, 152, 168, 169, 177	69, 70, 86, 105, 106, 113, 114, 126, 133, 149, 152, 168, 169, 177	55, 89	176	
150	150	110, 116, 117, 118* 119	118*	
72, 73, 171, 172	72, 73, 171, 172			
85, 88, 148, 170, 172	58, 60, 83, 84, 85, 93, 94, 137, 138, 141, 142, 143, 148, 172, 174	85, 148, 159, 160, 161, 164, 165, 166, 170	120, 139, 140	88, 109
58	54*, 56, 57, 58, 112, 131	144	54*	

BEDSTRAWS   (Rubiaceae)

Here are four weeds easy to identify as members of a group. Common characteristics are their leaves — in whorls around the stem; the flowers — always with 4 petals, in clusters and springing from the leaf axils; the round green fruits — hairy or with hooked bristles, having the habit of sticking to anything that touches them. All are creeping weeds forming a mat and sometimes a dense, smothering tangle relentlessly clutching plant victims and dragging them down to the dust. The rootstock supporting this abundant growth is remarkably tenuous, only a few spreading hairs.

CROSSWORT, *Galium cruciata*, is a perennial European with tiny yellow flowers clustering above the 4–leaved whorls. The stems are hairy and are 4–angled. Up to 1in (2.5cm) long, the oval-elliptical 3–veined leaves are largest in the middle of the stem, hairy on both sides, and yellowish-green. The Crosswort flowers in May and June and ranges throughout Britain as far north as Moray and Inverness, west to the Inner Hebrides and Ireland; in Europe northwards to the Netherlands, east to Brandenburg and southern Poland and across to Siberia.

Galium aparine has many common names, and GOOSEGRASS (geese feed on it greedily) or CLEAVERS, Hairif, Sticky Willie, Robin-run-the-hedge, Beggar's Lice, Scratch Grass and, also in America, CATCHWEED BEDSTRAW, all add up to the commonest of the bedstraws, with running stems up to 5ft (150cm) long, and downward-directed prickles by which it claws its way up and over other plants. Though its troublesome habits put it in the front rank of garden weeds, it is at least an indicator of good loam, for although it will grow in hedges and waste places it prefers a cultivated, deep loam or clay soil rich in nutrients and containing humus. The lanceolate leaves, 6–8 in a whorl, have a single central vein, and the spines of the margins point backwards. The flowers are sparse, only 2–5 of them

CROSSWORT

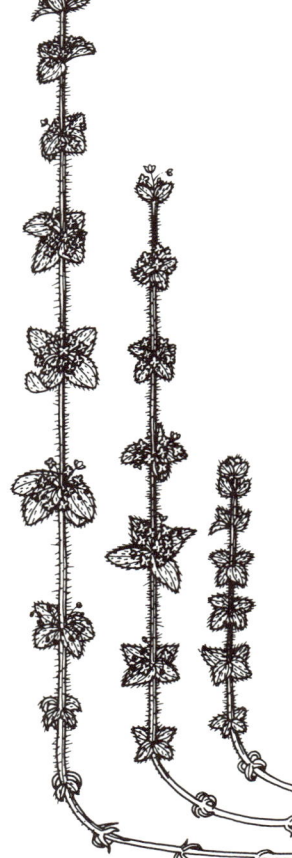

together instead of the thicker clusters of the other bedstraws, but each plant yields some 300–400 seeds. An annual or biennial weed, its flowering period is from May to November.

Goosegrass is distributed all over Europe and northern and western Asia, throughout the United States and north into Canada from Newfoundland to British Columbia and Alaska.

HEATH BEDSTRAW, *Galium saxatile*, is a small mat-forming perennial not to be found on lime or chalk: as its name indicates, heaths are its homeland, meaning an acid soil. It has numerous prostrate branches of narrow oval leaves in whorls of 6–8, with the margin-bristles pointing obliquely forwards. The flowering shoots lie along the ground and rise up at the ends in panicles of tiny white flowers in July and August. The plant turns black when dried. Common in the British Isles, it ranges across western Europe from northern Spain to southern Scandinavia, eastwards to Bohemia, and west across the Atlantic to Newfoundland.

Lastly we look at LADY'S BEDSTRAW, *Galium verum*, called YELLOW BEDSTRAW in North America where it is often used as an ornamental. It is a perennial flowering from July to September and varying in height from 4in to 2ft (10–60cm). The leaves, single-veined and almost threadlike, vary in number from 8 to 12 in a whorl and have their margins rolled under. They are dark green and rough above, often downy underneath. The small yellow flowers, in conspicuous panicles branching from the leaf axils, smell of honey when fresh. Dried, they have the hay-like smell of coumarin, which yields an anti-coagulant. Oddly, the flowers of Lady's Bedstraw also have a substance which acts in the opposite way, being of proven value as a styptic and as rennet for curdling milk into junket and cheese. In the Scottish Highlands the roots were used in conjunction with alum to dye red. The name Lady's Bedstraw alludes to the Virgin Mary's giving birth to her son in a stable: straw, grass or herbs were used for the poorer sort of bedding even in the last century. Common in the British Isles, the plant ranges throughout western Europe and western Asia. In North America it is an occasional weed from Newfoundland to Ontario and the Dakotas. It was first introduced into the United States in east Massachusetts.

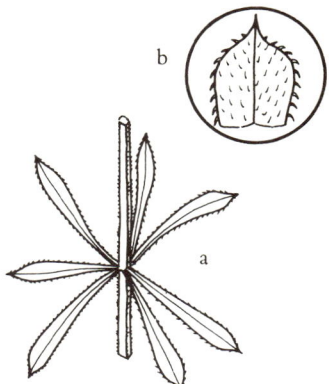

GOOSEGRASS or CLEAVERS *a*. Whorl of leaves *b*. Leaf-tip enlarged

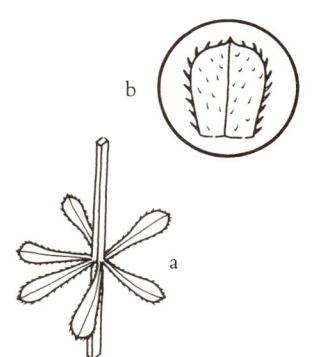

HEATH BEDSTRAW *a*. Whorl of leaves *b*. Leaf-tip enlarged

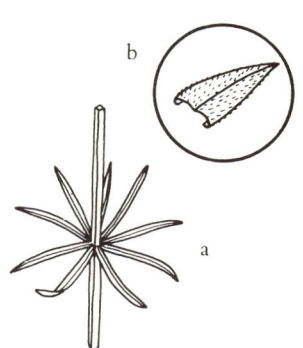

LADY'S BEDSTRAW or YELLOW BEDSTRAW *a*. Whorl of leaves *b*. Leaf-tip enlarged

54

BELLFLOWER
(Campanulaceae)

CREEPING CAMPANULA or
BELLFLOWER

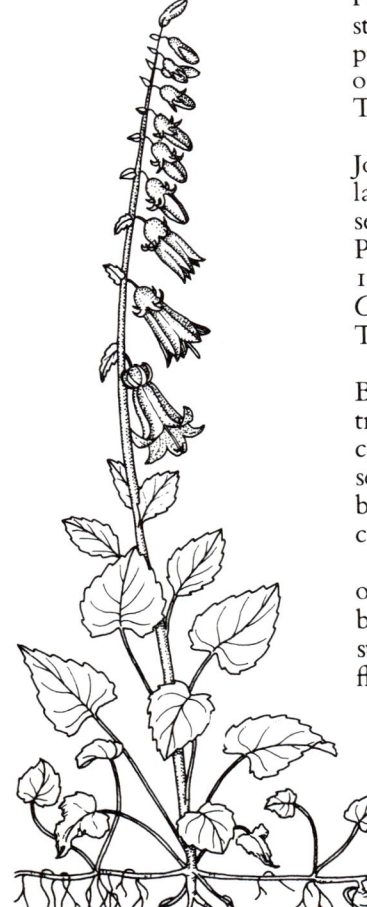

CREEPING CAMPANULA or BELLFLOWER, *Campanula rapunculoides*, is an example of a garden plant that has become a weed, and a weed that is hard to part with — in more ways than one. It has an ethereal charm, slender spires of nodding lavender-blue helmets shading up into buds of palest green. But — cherish it and you have a problem on your hands: it is almost ineradicable. From its stoloniferous roots, rhizomes or underground stems creep in every direction, break into shoots and produce new plants. Added to this, the potential offspring from a single inflorescence average about 400. The seeds are scattered by the wind.

An introduction into Britain, it was being grown by John Tradescant in his garden in 1633. Two centuries later a Dr Fisher found it in the dry hills of Tauria and sent seeds to William Anderson, curator of the Chelsea Physic Garden, who successfully raised them, in July 1823 writing a description of the beautiful plant for *Curtis's Botanical Magazine*. It was then called the Taurian Bell-flower.

Introduced into North America as a garden plant the Bellflower escaped into the wild to become a troublesome coast-to-coast weed. As it is almost completely impervious even to repeated applications of selective herbicides, the only way to banish it (if you can bear to do this) is by groping after its roots and keeping constant vigilance thereafter.

A perennial, it grows up to 3ft (90cm) tall. The spikes of blue flowers rise from rosettes of heart-shaped leaves, but the alternate and short-stalked leaves on the flower-stems are oval to lanceolate. The Creeping Campanula flowers from June to September or later.

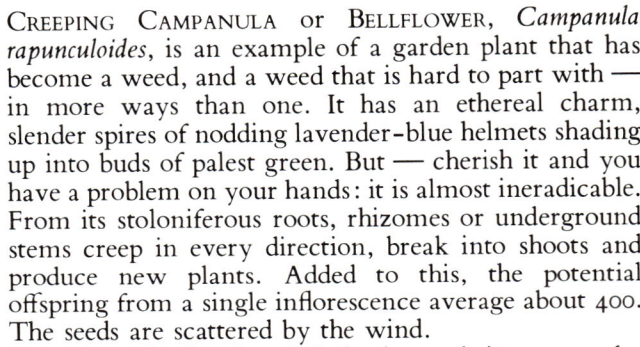

BIDENS   (Compositae)

The genus *Bidens* numbers about 200 species, most of them American. They belong to the largest family of flowering plants, the Compositae, all of which have a daisy-like flower which is in fact a composition (hence the name) of two kinds of flowers on the one head: tubular ones in the centre, called disk-florets; and strap-shaped ones radiating outwards, called ray-florets. The appearance can be tufted, as in thistles, and this applies to our two species of *Bidens*. Their single-seeded fruits have hooked bristles.

In *Bidens frondosa* the fruit has two barbed horns, giving the plant its common name of DEVIL'S BEGGARTICKS. This is an annual weed with a shallow much-branched tap-root from which the stems grow to a height varying from 8in to 4ft (20–120cm). The leaves are opposite on the stem and are compound with 3–5 lance-shaped leaflets. The disk flowers are orange, the ray flowers orange-yellow. Devil's Beggarticks ranges throughout the United States and north into Canada from eastern Quebec to western Ontario.

It has invaded Britain, where BEGGAR-TICKS and Stick-tight are its common names, and it is now established locally in wet places and waste ground.

So far, SPANISH NEEDLES, *Bidens bipinnata*, remains an American weed, but even in the United States is found only in the eastern half below northern New England and the Great Lakes states. It is an annual with a much-branched tap-root, and the name, Spanish Needles, comes from the long seed-vessel looking like a trident but with 4 barbed prongs. Before they are detached they are bunched in the calyx like arrows in a quiver. The leaves are opposite and very much divided, having up to 9 leaflets and each of these lobed or divided. The word pinnate comes from the Latin *pinna*, a feather. Botanically it means a leaf with leaflets arranged feather-fashion. A leaf is bipinnate when the leaflets themselves are thus arranged. Hence the botanical description of the leaves of Spanish Needles.

DEVIL'S BEGGARTICKS *a*. Its fruit
b. Fruit of SPANISH NEEDLES

BINDWEEDS
(Convolvulaceae)

These embrace (and this is an apt word) a most serious and troublesome set of weeds. They are stranglers and smotherers, their tap-roots going deep down and producing numerous lateral creeping roots that form new plants at the tips. They flower all summer and seed profusely. If you can get rid of bindweeds you are indeed a good gardener.

The BELLBINE or LARGER BINDWEED, *Calystegia sepium*, has been decorating English hedgerows and plaguing English gardeners for 500 years. William Turner lists it in his *Names of Herbs* (1548) as a British plant. It is common in the south of England and increasingly rare northwards. America also knows it as *Convolvulus sepium*, the HEDGE BINDWEED. The names Hedge Lily and Old Man's Nightcap are tributes to the beautiful white (sometimes pink) trumpet flowers nearly 2in (5cm) across, though country folk could not find names bad enough to call it. Devil's Garter, Devil's Vine and Hellweed refer to its pestilential character and the difficulty of getting rid of it. Bearbine and Ropewind tell of its climbing habit, Woodbine of its twining in a spiral manner, binding branch to branch.

A vigorous climber, the Bellbine is able to reach the top of a 20-ft tree. Clambering over a shrub the dense growth of alternate arrow-shaped leaves can almost smother it. Yet for beauty's sake I allow one specimen of it to grow unmolested. It climbs up a tall box bush by the edge of a pond where in August its great white trumpets shine against the dark green foliage, and before the frost comes each leaf is a golden heart against it. But if it dared to envelope a more precious shrub I would forget it as Hedge Lily and think of it only as Hell Weed. It flowers in England from June to August. The eastern half of the United States, and southern Canada, know the Hedge Bindweed well as a nuisance with a longer flowering time, mid-May to September.

A closely allied species is the European *Calystegia silvatica*, called the GREAT BINDWEED, with white or pink flowers usually larger, about $2\frac{1}{4}$–3in (5–7cm) across and

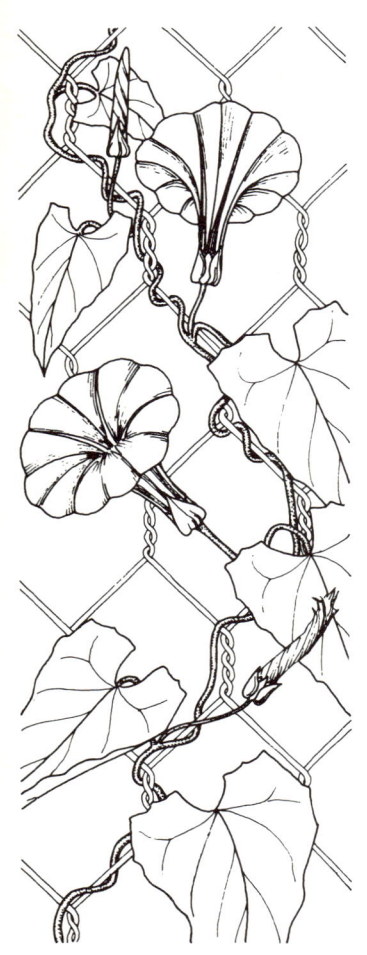

BELLBINE or HEDGE BINDWEED

with larger foliage, often of a darker and bluer green. The distinguishing difference between it and *Calystegia sepium* is in the bracteoles, the leafy structures at the back of the flowers. Whereas in *Calystegia sepium* they are not or scarcely inflated, not overlapping and therefore not completely hiding the calyx, in *Calystegia silvatica* they are strongly inflated and completely conceal the calyx. *Calystegia silvatica* is not native to Britain but has been there since 1777 at least. It too flowers from June to August.

BINDWEED, *Convolvulus arvensis*, known as FIELD BINDWEED in North America, is one of the thirty weeds in the United States described by the American weed expert M.A. McCall as truly noxious in every sense of the word. He drew up a table of ten characteristics, each one of which was enough to label a plant a weed. Bindweed answered to all ten. Charming to look at with its pink-and-white striped funnels, the Bindweed (also known as Cornbine, Lesser Bindweed, Barbine, Corn Bind and Creeping Jenny) is a perennial whose underground stems can cover an area of 30 square yards in a season. Its roots have been known to penetrate down 23ft (nearly 7m). A single plant can produce about 550 seeds, so prevent seeding by picking off the flowers, and starve it out by continually pulling away the foliage and trowelling out the roots. If you are more persistent than the Bindweed (also called Devil's Guts) it will gradually give up trying.

The stems climb by twisting in a counter-clockwise direction round the stems of other plants. The leaves, alternate and less than 1–2in (2–5cm) long, are arrow-shaped. The inch-long (2.5cm) flowers, which have nectar and are scented, grow on long stalks from the leaf axils. The bracteoles do not overlap the calyx. Distribution is throughout the temperate regions of both east and west hemispheres.

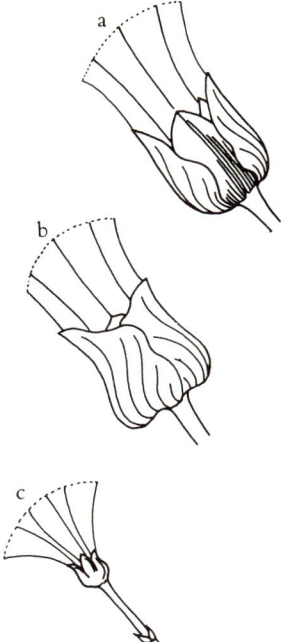

Calyx and bracteoles
a. BELLBINE or HEDGE BINDWEED
b. GREAT BINDWEED
c. BINDWEED or FIELD BINDWEED

BINDWEED or FIELD BINDWEED

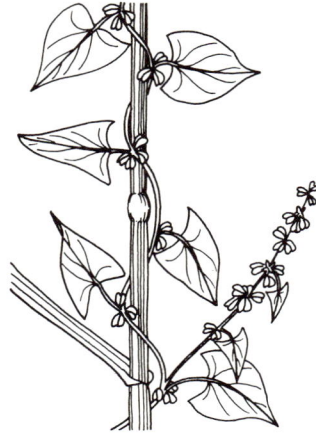

BLACK BINDWEED

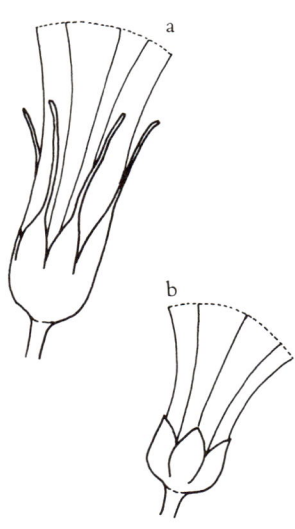

Calyx of *a*. IVY-LEAF MORNING GLORY *b*. TALL or COMMON MORNING GLORY

Polygonum convolvulus does not belong to the Convolvulus family but we include it here because it behaves like and looks like a bindweed, and indeed is called BLACK BINDWEED. It is common throughout the British Isles and Europe whence it was introduced into North America, to be distinguished from the true Bindweeds by its sprawling habit of stems (often coloured red) radiating from the rootstock, and by its flowers which instead of being funnel in form are inconspicuous whitish-green and short-stalked, 2–6 in a lax spike. But again the leaves, nearly smooth above and mealy beneath, are alternate and arrow-shaped. The flowering period is from July to October and each plant produces up to 200 seeds. Although it will grow in waste places such as rubbish dumps, the Black Bindweed prefers a nutrient-rich soil and moderately acid loams.

The IVY-LEAF MORNING GLORY, *Ipomoea hederacea*, is a Convolvulus that torments American gardens but spares European ones. An annual, its leaves are alternate but different in shape, being 3-lobed and occasionally 5-lobed, 2–5in (5–12cm) long. Its funnel-shaped flowers are pale to sky-blue when fresh, quickly changing to rose-purple. The long narrow sepals enclosing the white tube are densely hairy or bristly. It is an invader from tropical America spreading throughout all the eastern and south-eastern, central and south-western United States and north into Ontario.

The TALL or COMMON MORNING GLORY, *Ipomoea purpurea*, is another invader from tropical America, regarded in the United States as a noxious weed, but treasured by British gardeners who love to grow it for its heavenly-blue trumpets. Botanic gardens devote glass-house space to exhibiting the range of its colours: flowers on a single plant can be blue, purple, red, white, or variegated, and up to nearly 3in (7cm) long. In this plant the sepals are oval, the leaves again heart-shaped. An annual, it blooms from July to September.

BLACKBERRY or BRAMBLE (Rosaceae)

People who study brambles are nicknamed 'batologists', and one can understand why! There are some 400 species in Europe, over 300 in the United States and botanists are still trying to classify them. The best they can do so far in Britain is to designate the BLACKBERRY or BRAMBLE as '*Rubus fruticosus* aggregate'.

This is the delicious black fruit we gather from hedgerows (in England's Suffolk before 11 October, for then 'the devil gets into them') to make jelly and jam. Hedges are their hiding places, garden hedges included, and long prickly arms soon reach out to damage young lettuces or whatever other plant is in their path. They must be dug out as soon as discovered or they will leapfrog everywhere, each growing-tip rooting and becoming a new plant.

An aggressive enemy, the Blackberry is able to reproduce itself without fertilization in a process called apomixis. The 5–petalled flowers in clusters vary from white to pink, and the alternate leaves can be single or with three or five leaflets, with or without down above, green or bluish underneath. The stems are perennial or biennial, climbing or crawling along the ground and with few or many prickles, deciduous or evergreen. A variable plant indeed. What does not vary is its ability to turn any piece of land into an impenetrable thicket.

This variability applies also to the ALLEGHENY BLACKBERRY of the United States, *Rubus allegheniensis*, but it has two distinguishing features: its flowers are usually white, rarely pink or pinkish, and its fruits come dark-purple, or red, white, or yellow.

BLACKBERRY or BRAMBLE
a. Flowers *b*. Fruit *c*. Stems rooting at tips

BRYONIES (Dioscoreaceae and Cucurbitaceae)

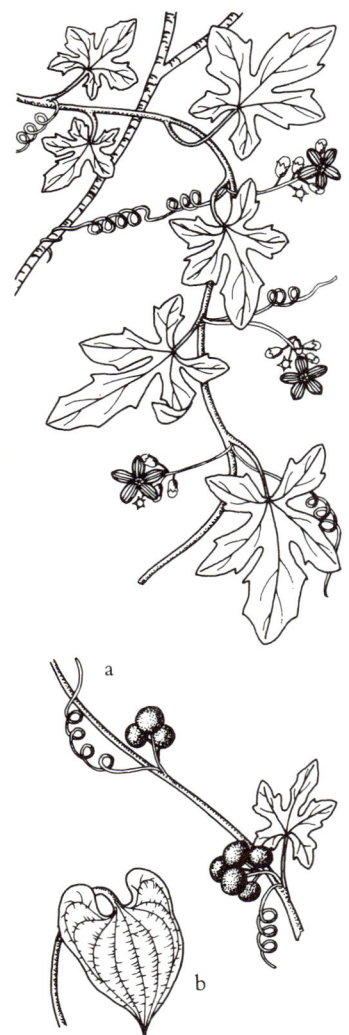

The BLACK BRYONY, *Tamus communis*, is a climber of rapid growth confined to Britain in Wales and England as far as south Cumbria and Northumberland, and to southern and western Europe east to the Black Sea. It climbs by twining round a support in a counter-clockwise direction and is easily recognized by its shining heart-shaped alternate leaves with a tapering point, bronze in the autumn. Blooming in spring and early summer, the flowers are small and yellowish-green, the male flowers on slender spikes borne on one plant, the female flowers in much shorter and closer racemes on another plant. The berries are bright red. A perennial belonging to the Dioscoreaceae family, the Black Bryony grows from an oval blackish tuber.

The WHITE BRYONY, *Bryonia dioica*, belongs to the Cucurbitaceae family, claiming relationship with cucumbers, melons, gourds, pumpkins, vegetable marrows and squashes. Its method of climbing is different. A perennial, its new spring shoots covered with long glistening hairs trail off in every direction until one of them meets a support up which it can start its climb by means of wonderful tendrils. The tip of the tendril twines around the nearest twig. The rest of the tendril is then tightened into a spiral and the trailing stem drawn up behind it. To avoid self-twisting, the coiled tendril straightens itself a little way in the middle, from there coiling in the opposite direction. It was Charles Darwin who observed this marvellous device one night during a gale, watching the Bryony's elastic springs giving to the wind and re-coiling. Another climber not so endowed would have been torn from the tree.

The alternate leaves are 5–lobed, the small clusters of beautiful little 5–petalled flowers pale greenish-white with yellow stamens. The berries are red or orange with 3–6 large fat seeds, yellow and mottled with black, or black mottled with yellow. The White Bryony is common in England and Wales, central and southern Europe.

WHITE BRYONY *a*. Its fruit
b. Leaf of BLACK BRYONY

BUTTERCUPS
(Ranunculaceae)

How innocent look buttercups on a May morning! But for all the beauty of their golden waxy flowers the acrid juice they contain can be fatal to cattle browsing among them, particularly in two of the common species, though when dried and made into hay they become harmless.

Gardenwise, buttercups are a double menace. Liking a soil rich in minerals, they deprive border plants of food they need. Secondly, they exude a substance which retards and stunts the growth of neighbouring plants. All are perennials. The following three are the most common.

The MEADOW BUTTERCUP, *Ranunculus acris*, called the TALL BUTTERCUP in the United States, attains a height of 6–40in (15–100cm) and has 3–lobed stem-leaves, alternate and without stalks. The roots are thick and fibrous, and the achenes (single-seeded fruits) have a nearly straight beak. The solitary flowers, usually bright yellow, are sometimes paler or even white. Normally they have 5 petals. The Meadow or Tall Buttercup is found throughout the whole of the United States excepting an area between central Montana and eastern Minnesota. It is common in Britain and has a wide distribution throughout temperate and even arctic Europe and Asia.

The CREEPING BUTTERCUP, *Ranunculus repens*, has long stout roots and leafy runners which root at the nodes, the point on the stem where the leaves arise. The flowering stems, 6–24in (15–60cm) tall, leafy and hairy, bear bright-yellow 5–petalled solitary flowers from May to July. The leaves are 3–lobed, each lobe or leaflet being further divided into 3 segments deeply toothed. The middle lobe, which is long-stalked, often forms 3 distinct leaflets of its own. The general appearance is a leaf triangular and sharp in outline. This buttercup seeds more sparsely, but it often has as many as 25 runners which can quickly make a large colony, in a single season spreading over an area of 40 square feet. A native

MEADOW or TALL BUTTERCUP

of Europe, Asia Minor, Siberia and North Africa, it is common across the northern half of the United States excepting an area between central Montana and eastern Minnesota, though it occupies a distinct area in the south central States. In Canada it is found from Newfoundland into Alaska.

CREEPING BUTTERCUP

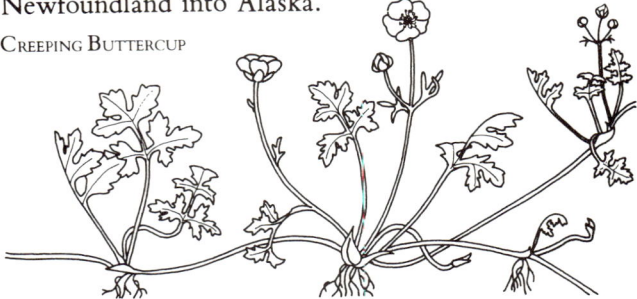

The BULBOUS BUTTERCUP, *Ranunculus bulbosus*, is thus named because of its tuberous base. Its height varies from 6 to 16in (15–40cm), and it too flowers from May to July. The stems are erect and hairy. The lower leaves are 3–lobed and alternate, the middle lobe being long-stalked, while the upper leaves grow close to the stem and are cut into narrow, often linear, segments. All the leaves are usually hairy. A distinguishing feature is that the 5 pale yellowish sepals are strongly reflexed, the flowers themselves having 5 glossy yellow petals and many stamens. The achene has a short beaked hook which tells that it is distributed by attaching itself to the fur of animals. It is common throughout Europe and was introduced into America.

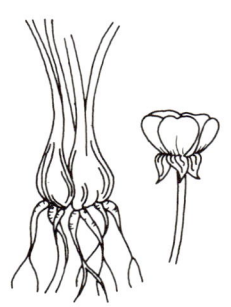

BULBOUS BUTTERCUP, flower and root

The LESSER CELANDINE, or PILEWORT, *Ranunculus ficaria*, is distinguished by a completely different leaf, heart-shaped and undivided. Both leaves and flowers grow singly from the root tubers, a collection of tiny bulbils which easily detach themselves and slip away into the soil, making the Lesser Celandine one of the most difficult weeds to get rid of, for each bulbil will grow into a new plant next year. The waxy golden flowers, which have 8–12 pointed petals, fade to white. Do not confuse this invasive weed with the larger-flowered fibrous-rooted variety from the Mediterranean, brought to England in the 17th century by John Tradescant, and a worthy garden plant. The Lesser Celandine is distributed throughout Great Britain and Ireland and the Channel Islands, Europe and western Asia. It has been introduced into North America.

LESSER CELANDINE or PILEWORT, flower, leaf and root

CAMPIONS
(Caryophyllaceae)

Know the Campions by the bulbous calyx below the white, red or pink petals. Summer is really with us when they are in bloom. It is then that the Silver Y moth, *Plusia gamma* whose brown forewings bear the gamma sign (γ) of the Greek alphabet, emerges from its chrysalis to feed upon their flowers.

The RED CAMPION, *Silene dioica*, is so much like the WHITE CAMPION, *Silene alba*, in America called the WHITE COCKLE under the synonym *Lychnis alba*, that it was until fairly recently regarded as a red-flowering variety. There are, however, several distinguishing features beside colour. The teeth of the Red Campion's seed capsules are revolute (rolled downwards): the 10 teeth of the White Campion's stand erect. The Red (*Lychnis dioica* in America) is a biennial to perennial with a slender creeping rootstock: the White is a short-lived perennial and sometimes annual or biennial with a thick, almost woody stock. The Red Campion produces numerous non-flowering decumbent stems that rise to about 8in (20cm), and erect flowering stems 12–36in (30–90cm): the non-flowering stems of the White Campion are few and short, the flowering stems 12–40in (30–100cm) long.

The larger flowers of the White Campion are slightly scented in the evening, the Red Campion is scentless; the Red Campion produces black seeds, the White has grey seeds. Both have 5 deeply cleft petals and lanceolate opposite leaves.

Naturalized from Europe, where it is common, the White Campion or White Cockle is found throughout the northern half of the United States and southern Canada from British Columbia to Nova Scotia. It is common also in the British Isles. The Red Campion is found occasionally in North America.

The BLADDER CAMPION, *Silene vulgaris*, in the U.S. *Silene cucubalus*, grows 12–18in (30–45cm) tall or taller, and the white flowers have 5 petals, again deeply cleft. It is a perennial with branching woody roots and elliptical to oval leaves that are opposite, the lowest being short-

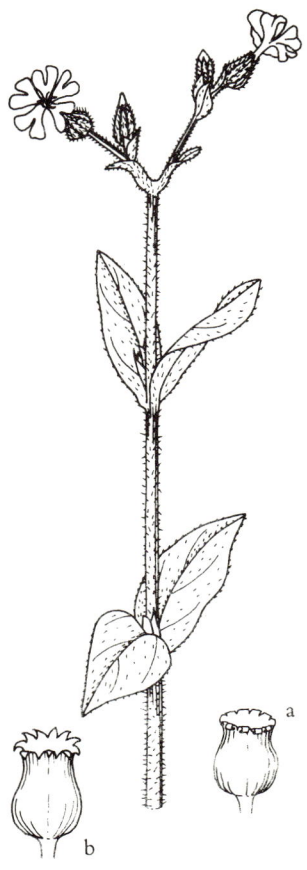

RED CAMPION *a*. Its capsule
b. Capsule of WHITE CAMPION
or WHITE COCKLE

a

b

Capsule of *a*. BLADDER CAMPION
b. NIGHT-FLOWERING CAMPION
or NIGHT-FLOWERING CATCHFLY

stalked, the upper ones growing close to the stem. The pale green or reddish bladder (calyx) has 20 veins with a strong connecting network. It encloses the capsule which has 6 erect teeth. Common in most of Great Britain and Ireland, Europe and temperate Asia, the Bladder Campion was introduced into the U.S. in ballast from Europe dumped in Philadelphia.

Silene noctiflora, the NIGHT-FLOWERING CAMPION, is an annual having the same 5 deeply cleft petals, pale pink above, yellowish beneath and scented at night. The calyx, a long woolly and sticky tube, becomes swollen when the fruit is formed, the capsule having 6 recurved teeth. From 6–24in (15–60cm) tall, the stout erect stems, hairy and sticky, grow from strong roots. The leaves, which are opposite, are roughly oval, narrowing towards the top of the stem. NIGHT-FLOWERING CATCHFLY is another name for it, both in Britain and in North America where it is generally distributed.

CARPET WEED (Aizoaceae)

The botanical name for the CARPET WEED is *Mollugo verticillata*, which means 'soft' and 'whorled'. It is a soft plant and everything about it is whorled — the basal leaves around the rootstock; the way the stems radiate from it; the leaves around the stems, and the tiny flowers that radiate from the leaf axils. Seen from above, the plant is a series of four wheels lying flat on the ground. It is a late-starting but quick-growing annual which rapidly covers any fertile soil, making mats in borders and lawns. The narrow leaves are in groups of 5 or 6, occasionally 3 and up to 8, in length $\frac{1}{2}$–1in (1–2.5cm). The flowers have no petals, but there are 5 long petal-like sepals white inside and oval in shape. It has a long flowering period, June to November, and produces many seeds.

A native of tropical America, the Carpet Weed has invaded almost the whole of the United States, and parts of Canada.

CARPET WEED

CHICKWEEDS
(Caryophyllaceae)

The Chickweeds are an interesting group. About 100 species are cosmopolitan, and delicate as they look and with roots penetrating only a few centimetres into the ground, they can yet tolerate a low temperature and subsist on little, though the better the soil the better they grow. They are mat-like spreaders and many are automatically self-pollinating, which means they do not have to rely on dry weather and insect visitors to bring them to fruition. Not a garden escapes them. Regarded as troublesome because they are somewhat resistant to most selective herbicides, they are at least easy to uproot. In the vegetable garden if they spring up among tender young pea plants they are best left alone, since pulling up their straggling surface roots will damage those of the young peas. On a light soil which dries up in summer they help to retain moisture.

CHICKWEED or COMMON CHICKWEED, *Stellaria media*, is one of the cosmopolitans, familiar throughout the world. It has travelled wherever white settlers have trod and is equally at home in the Arctic Circle and in South America. In Britain it goes by many different names: Chickwittles, Cluckenweed, Mischievous Jack, Skirt Buttons; in Ireland, Tongue Grass; in America, among others, Starweed, Winter Weed and Satin Flower. French-speaking Canada calls it Cyrillo.

Some weeds are unwholesome to look at. Not so the Chickweed with its fresh green, pointed oval leaves growing in opposite pairs up stems which have a single line of hairs. The clustered buds are conical on long stalks from the leaf axils, opening to small white flowers with 5 petals so deeply divided that they look like 10. They usually last only a day and have 3–8 red-purple stamens, the number being greater if the plant grows in the open in full light, fewer if it grows in shade.

Although it is an annual, the life-span of an individual Chickweed may extend over several months; and 5–7 weeks after germinating, its seeds can be ripened and shed. The plant goes on flowering throughout the year except in severe weather, and each produces 2,500–15,000 seeds. Since there can be three

CHICKWEED or COMMON CHICKWEED

generations in a year, the potential offspring over a period of 12 months could, on the lowest reckoning and if the majority were to survive, number over 15,000 million plants. Up to you to reduce this horrifying potential population!

Added to its fecundity, the Chickweed distributes itself in a variety of ways. Its seeds remain viable even after passing through the digestive tracts of birds and animals, and experiments have proved that after 90 days' immersion in sea-water they germinate, bloom and fruit. It is easy to see how by land and sea the Chickweed has spread throughout the world.

Leaf of STICKY MOUSE-EAR CHICKWEED

Of the eleven different Mouse-ear Chickweeds found in Britain alone, five are regarded as weeds, including one which is a garden plant that has become a weed. Commonest in Britain is the STICKY MOUSE-EAR CHICKWEED, *Cerastium glomeratum*, known in North America under its Latin synonym *Cerastium viscosum*. It is another cosmopolitan and can be distinguished by its pale yellowish-green leaves, broadly oval and opposite; its flowers growing in compact clusters; its 10 stamens and 5 styles; and hairy and very short flower-stalks. The 5 white petals are not so deeply cut. The flowering period is from April to September, the seeds mostly germinating in autumn or late summer. Growing to a height of 18in (45cm) it dies down in winter to a leafy rosette.

Leaf of COMMON or MOUSE-EAR CHICKWEED

The COMMON MOUSE-EAR CHICKWEED of Britain, *Cerastium holosteoides*, is the COMMON or MOUSE-EAR CHICKWEED of North America, *Cerastium vulgatum*. A perennial, it flowers from June to August and straggles along the ground in a mass of barren shoots, only pushing up the ones that flower. Again the 5 petals are cleft, the leaves opposite but dark greyish-green densely covered with white hairs, those on the flowering shoots elliptic to oval with no stalks, those on the barren shoots more blunt. Another cosmopolitan, it minds neither cold nor heights: in Scotland it reaches an altitude of nearly 4,000ft.

The FIELD MOUSE-EAR CHICKWEED, *Cerastium arvense*, in the United States and Canada called the FIELD CHICKWEED, is a perennial flowering from April to October, again a branched creeping weed but distinguished from the others by the lower leaves'

FIELD MOUSE-EAR CHICKWEED or FIELD CHICKWEED

commonly having axillary leaf clusters: that is, clusters of leaves growing from the joints of the usual opposite pairs of leaves. All the leaves are downy and narrow, the flowers disproportionately large, $\frac{1}{2}-\frac{3}{4}$in (12–18mm) across, the length of the leaves being $\frac{1}{4}-\frac{3}{4}$in (6–18mm). This Mouse-ear has the habit of rooting at the leaf-joints of the ground-shoots, resulting in an ever-spreading mat. It is a pest in lawns.

Finally we have the renegade *Cerastium tomentosum* which began life in gardens as an imported rock plant enchantingly called SNOW-IN-SUMMER and inevitably becomes a weed to be grubbed up by the handful lest its exuberant growth smother its neighbours. A native of south-east Europe and the Caucasus, it has narrow silvery leaves and in June covers itself with starry white solitary flowers having the usual 5 bifid petals, $\frac{3}{4}-1$in (18–25mm) across. It is not common in North America as a weed.

Leaf of SNOW-IN-SUMMER

CHICORY (Compositae)

Belonging to the Daisy family the vivid blue-rayed flowers of the CHICORY or SUCCORY, *Cichorium intybus*, are eye-catching beauty from June to October growing along roadsides. Chicory is not so welcome in a garden, however, particularly in a lawn, where its deep tap-root and rough, tough stems make it formidable in combat. But in a border these same tap-roots can be useful in pioneering the subsoil for other plants.

Chicory grows to a height of 1–4ft (30–120cm), much taller in cultivated ground, even up to 8ft (2.4m)! The lower leaves, long and toothed with a strong central rib, resemble those of the Dandelion. These branch from sheaths emerging from the rootstock, the leaves on the flower-stems being alternate and narrower, clasping the stem and smaller towards the top. The flowers open in the morning and close in the afternoon, but as each fades a new one takes its place.

A native of the Mediterranean region, Chicory is cultivated in Europe, the dried and ground roots yielding the chicory of commerce as a substitute for or additive to coffee. The roots are boiled and eaten by the Arabs, and it is from their word *chicouyeh* that the English common name is derived. The Romans used it as a salad, the herbalists for rheumatic and liver complaints.

Chicory is distributed throughout Europe, most of the eastern, central and western states of North America, and north into southern Canada from Nova Scotia to British Columbia.

CHICORY

CHRYSANTHEMUMS
(Compositae)

The botanical name for the first weed in this group is *Chrysanthemum parthenium*, from the Greek *chrysos*, gold, and *anthos*, a flower; the specific epithet, as Plutarch recounts, commemorating the cure of a man who fell from a height during the building of the Parthenon in Athens. Another derivation could be from the Latin word *parthenice*, meaning a kind of herb. Herb it is, for FEVERFEW was a febrifuge used for curing giddiness and 'driving awaie agues'. It is a fibrous-rooted perennial strongly aromatic like the cultivated chrysanthemum and growing 10–24in (25–60cm) tall, with erect and somewhat downy stems branched above in corymbs, showing the flowers all at the same level. These, blooming in July and August, are daisy flowers of white ray-florets, short and broad, and yellow disk-florets in the centre. The leaves, 1–3in (2.5–8cm) long, are yellowish-green and alternate, the lower ones being long-stalked and divided into toothed or lobed leaflets making roughly a heart-shaped outline, the upper leaves shorter in the stalk and less divided.

In North America the Feverfew grows to a height of 3ft (90cm). It is an introduction, as it is in Britain, its homeland being south-eastern Europe, Asia Minor and the Caucasus. It is now established in South America. The golden-leaved variety *aureum* is still popular as a foliage-plant for bedding-out, but it too can become a nuisance.

Chrysanthemum leucanthemum has several common names: in Britain it is the MARGUERITE, MOON DAISY, DOG DAISY or OX-EYE DAISY, also in the United States the FIELD OX-EYE DAISY. A tall-growing perennial, its height is 8–28in (20–70cm), the flowers also being larger and with longer white ray-florets, solitary on long stalks and in bloom from June to August. All the leaves are dark green and alternate, the lower ones on the stem being roundish to almost spatulate, toothed and on long stalks; the upper stem-leaves oblong, blunt, toothed or divided and clasping the stem.

The habit of the plant is a graceful looseness. Childhood memories are of banks of them in long grass,

FEVERFEW

MARGUERITE or OX-EYE DAISY
a. Flower *b.* Stem-leaf

swinging to and fro in the summer air. Naturalized from Europe, it inhabits most of the United States, Canada from Labrador to British Columbia, the British Isles and continental Europe to Lapland and Siberia.

The TANSY, *Chrysanthemum vulgare* (in North America *Tanacetum vulgare*, the COMMON TANSY), is a strong-smelling perennial with creeping underground stems (stolons) which send up new shoots. Its stiffly erect, leafy stems grow 12–40in (30–100cm) tall, and are tough, angled and usually reddish. The dark-green pinnate leaves having up to 12 pairs of sharply-toothed leaflets are a distinguishing feature of this Chrys-anthemum, as are the golden buttons of the flowers which bloom from July to September. Once cultivated as a medicinal and pot-herb, the Tansy ranges throughout Britain and in Europe as far east as the Caucasus, Armenia and Siberia. In the United States an escape from gardens, the Tansy has returned from the wild to trouble them.

TANSY or COMMON TANSY
a. Part of inflorescence
b. Stem-leaf

CINQUEFOILS (Rosaceae)

These are charming members of the Rose family and cousins of the strawberry, which they resemble in leaf and flower. They are widely distributed perennials. The name Cinquefoil refers to the 5-lobed leaf (though some species have only 3 and one bothersome member up to 12 pairs plus), and all have 5-petalled flowers.

The CREEPING CINQUEFOIL, *Potentilla reptans*, is one having 5 lobes spread like the fingers of a hand, and like the strawberry it produces radiating runners which quickly root at the nodes or leaf-joints. In this way it can colonize at the rate of more than 12 square yards in a season. The stock is thick and branched and both the

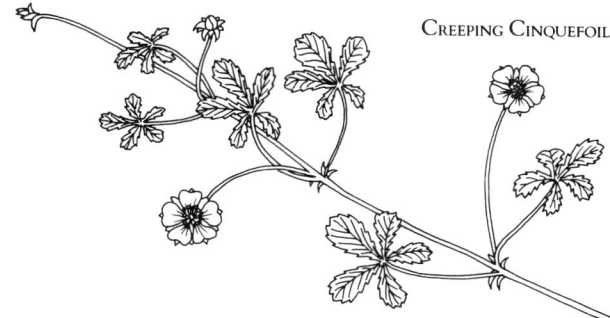

CREEPING CINQUEFOIL

leaves and golden flowers grow singly on long stalks
from the nodes of stipules, which are leaf-like
appendages growing in pairs. With a main tap-root,
blackish in colour, that penetrates down more than 12in
(30cm), the exuberance of its growth, and fruits that
contain an average of 90 seeds but up to 200, the
Creeping Cinquefoil is a weed to be dealt with sternly.
It is common throughout the British Isles and Europe
and has invaded both Americas.

Similarly widespread is the SILVERWEED, *Potentilla
anserina*. Do not be lulled by the charm of its silky
silvery leaves with their 7–12 pairs of main leaflets
alternating with smaller ones, and the bright yellow
solitary flowers poised on scarlet stems: it will carpet
your lawn if you let it, creeping, rooting and flowering
from long stolons, self-pollinating if insects fail to visit,
each head producing up to 50 fruits, the seeds of which
are loved by birds and thus distributed. The seeds also
travel by mud and rainwash. All but gardeners love this
weed which goes by other pretty names: Fern-
buttercup, Prince's Feathers, Midsummer Silver, Silver
Fern and Traveller's Ease. It flowers in Britain in July
and August, a month earlier and later in the United
States.

Leaf of SILVERWEED

The TRAILING TORMENTIL, *Potentilla anglica*, is a dainty
member of the family, with golden flowers, 4– or
5–petalled, only ¼in (6–18mm) across. The leaves that
form a rosette at the base are palmate with 5 leaflets, the
upper leaves having 3 leaflets, all the leaflets being
toothed at the top. Around the stem they grow singly,
or in twos, or whorled. The Trailing Tormentil flowers
from June to September. A European, it is naturalized in
south-west Nova Scotia and eastern Pennsylvania.

Leaves of TRAILING TORMENTIL

CLOVERS (Papilionaceae)

This family is recognizable by the pea flowers tightly bunched in heads and developing into pods, and by the trifoliate leaves. The 5-petalled flowers are botanically described as papilionate from their fancied resemblance to a butterfly, having a broad, erect petal called the standard and two lateral ones called wings, below which two lower petals unite in a single one called the keel. The leaves are alternate on the stems.

A creeping perennial rooting at the nodes, the WHITE or DUTCH CLOVER, *Trifolium repens*, has two distinguishing features: first, of course, the colour of its fragrant flowers, white tinged underneath with pink; and secondly the dark patch at the base of each rounded leaflet bordered by a white band. The oblong pod may be single-seeded but more usually has 4 seeds. Flowering from May to October, the White Clover is a fodder plant commonly escaping into gardens. It is found throughout the British Isles and continental Europe, North and South America.

WHITE or DUTCH CLOVER

The RED CLOVER, *Trifolium pratense*, is usually perennial, though lasting only a few years, with hairy stems 1–2ft (30–60cm) long, lying along the ground or nearly erect. The flower-head is visibly held in a cup of stipules and is pinkish-red or pink-purple, occasionally whitish, each head being up to 1¼in (3cm) across. Red Clover flowers from May to September.

The Red Clover is distinguishable from the ZIGZAG CLOVER (*Trifolium medium*) by the fact that its flower is not so dense in colour, and the 3-pointed leaflets have a white spot or inverted V across them; and from the CRIMSON CLOVER (*Trifolium incarnatum*) by the latter's having a dense flower-head, long and oval in shape. The Red Clover is distributed throughout the British Isles and continental Europe, North and South America.

BLACK MEDICK, *Medicago lupulina*, is a pest in lawns. The tough stems radiating from the rootstock close to the ground quickly smother the grass, and each stem can reach out to a length of 1–2ft (30–60cm). This species is recognizable by the tiny flower-head's being oval in shape and producing small single-seeded pods which turn black when ripe. It makes up for this limited seed supply and for the shortness of its life (it is an annual) by flowering the whole season in Europe from May to August, in the United States from March to September and even December. The Black Medick extends throughout the European continent and central and Russian Asia, and is widespread in North America.

Leaf of RED CLOVER

BLACK MEDICK

The SUCKLING CLOVER or LESSER YELLOW TREFOIL, *Trifolium dubium*, called the LEAST HOP-CLOVER in America, flowers from June to August. It is very much like the Black Medick but distinguished from it by having light brown seeds and paler round flower-heads instead of oval ones. These grow from the leaf-axils and turn dark brown. The standard is narrow, remaining as the pod develops and folding over it.

The Suckling Clover is a European but is common in some parts of Nova Scotia and the eastern and Pacific sides of Canada and the United States.

SUCKLING CLOVER or LEAST HOP-CLOVER

COLTSFOOT (Compositae)

Harbinger of spring is the COLTSFOOT, *Tussilago farfara*, with its pale yellow daisy-flower braving the March wind and appearing long before its rosette of heart-shaped toothed leaves. The flower-stems, which grow about 6in (15cm) high, have erect pink scales growing alternately and covered with a loose white cotton. The leaves, 4–5in (10–12.5 cm) broad, are also felted with this loose cottony wool, at maturity only underneath, and this used to be collected for tinder. The true stem runs underground, branching in every direction and making the Coltsfoot one of the most troublesome of weeds in the garden. The generic name of the plant comes from the Latin *tussis*, a cough. Bitter and astringent, the dried leaves used to be smoked as a remedy for asthma and lung complaints. The Coltsfoot is a perennial extending throughout the British Isles, Europe and North America.

COLTSFOOT, flowering plant and leaf

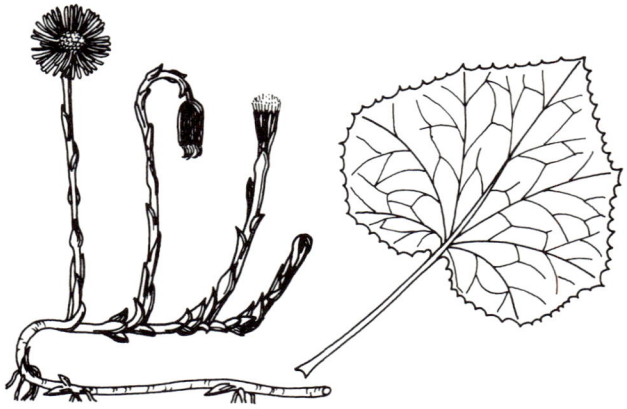

COW PARSNIP (Umbelliferae)

The COW PARSNIP, *Heracleum sphondylium*, is an umbellifer: that is, a plant with its inflorescence arranged like an umbrella, the flower-stalks coming like spokes from a single axis. Other names are Hogweed, Eltrot, Heltrot, and Keck. A biennial, it can grow to a great size, its hollow, hairy and channelled stems sometimes attaining a height of 7ft (2m). The leaves are 6in to 2ft (15–60cm) across and are rough, lobed in segments broadly oval, and toothed. The conspicuous white flowers, in bloom from June to September, have 5 deeply notched petals, those of the outer flowers being larger. The fruits are green withering to light brown. The Hogweed is a tough customer to deal with, its thick parsnip root growing deeply down. It is generally distributed throughout the British Isles and Europe, as well as North America. Strength characterizes its whole appearance: not surprisingly, its generic name comes from the god Hercules.

COW PARSNIP *a*. Part of fruiting umbel *b*. Flower *c*. Stem-leaf

CRESSES (Cruciferae)

CREEPING YELLOW CRESS

The Crucifers are identified by the cross made by the 4 petals of their flowers, hence the family name. Within this family are the Cresses, with pods growing up the stem, and the clustered flowers at the top still coming into bloom, giving the plant a rather ragged appearance. Pods and stem-leaves are alternate, the leaves more or less heart-shaped, in some species being compound, in others simple.

In a survey carried out recently by the Botanical Society of the British Isles CREEPING YELLOW CRESS, *Rorippa sylvestris*, was named as the most frequent garden weed in Britain. The flowers, about $\frac{1}{4}$ in (6mm) across, are yellow and bloom from June to September. A perennial and stoloniferous, the plant grows to a height of 9–18in (22–45cm), according to whether its branched stems are reclining or erect: it has both habits. The lower leaves are pinnate, that is divided into leaflets, and stalked. The upper leaves, much smaller and less divided, grow close to the stem with no stalks. Once established, Creeping Yellow Cress is extremely difficult to get rid of, not only because the underground stems root easily at the nodes but because, even if the plant is dug out, a new crop can spring up in a few weeks from roots left in the soil that may be no thicker than a thread of cotton. Although associated with river banks and wet places, Creeping Yellow Cress is equally at home in a sandy soil, and it can endure long periods of drought. It sometimes sneaks into gardens as an impurity among grass seed. Distribution is across Europe, all the Canadian provinces, southward in the United States to West Virginia and westward to Illinois and the coast.

The CUCKOO FLOWER or LADY'S SMOCK, *Cardamine pratensis*, is a meadow and pasture weed that has found its way into gardens in the British Isles, Europe and North America. But it is a charming one with sweet lilac flowers and sometimes white ones. Shakespeare knew them as 'Lady's Smocks all silver white' which with 'Cuckoo's buds of yellow hue do paint the meadows with delight.' The stem is erect, simple or branched, and nearly 1ft (30cm) high. The pinnate lower leaves have oval or rounded leaflets and form a rosette, the pinnate stem-leaves having linear leaflets. The pods, more than 1in (2.5cm) long, are supported on long stalks standing out from the stem at an acute angle. A perennial with a short rootstock often having tubers and occasionally stolons, the Cuckoo Flower blooms from April to June. Its generic name *Cardamine* means 'little cress'.

CUCKOO FLOWER

HAIRY BITTER CRESS, *Cardamine hirsuta*, is the sort of weed one takes little notice of, thinking to return to it when there is time to spare. It is small, about 6in (15cm) in height (though it can grow to a foot tall) with insignificant white flowers; and although an annual with a short life-span, its winter rosette of pinnate leaves, dark green with irregular round leaflets, can survive severe frosts and it can flower as soon as the temperature is favourable. In the course of a year several generations may mature their fruits and scatter the seeds far and wide.

This the Hairy Bitter Cress does by the aid of a marvellous ballistic mechanism. Each of the pods, which overtop the inflorescence, consists of two boat-shaped valves separated by a papery partition, and as the fruit ripens the tissues dry and contract, setting up considerable tension. When a ripe pod is touched, even by a sudden gust of wind, the valves tear away from the partition and coil back so abruptly and violently that the seeds are shot into the air as far away as 3ft. Though the annual seed production per plant averages 600, a large specimen can yield 50,000 seeds, and as a further aid to distribution the seeds become sticky when moistened and easily adhere to boots and birds' feet. Not surprising that the weed is common in the northern hemisphere.

HAIRY BITTER CRESS

WOOD BITTER CRESS, *Cardamine flexuosa*, is also called GREATER BITTER CRESS, though it grows only 4–20in (10–50cm) high. It is usually perennial and flowers from April to September. With a rosette of leaves at the base, stalked and pinnate with 5 or more pairs of rounded leaflets, it looks very much like the Hairy Bitter Cress but unlike it has a branched stem with alternate leaves growing up it, these also pinnate but short-stalked or clasping the stem. The flowers are white and insignificant, followed by pods not overtopping the inflorescence and more or less erect on slender stalks.

WOOD BITTER CRESS

Common in moist shady places, often growing by streams, the Wood Bitter Cress is a European that has found its way to the Far East, Newfoundland and Quebec.

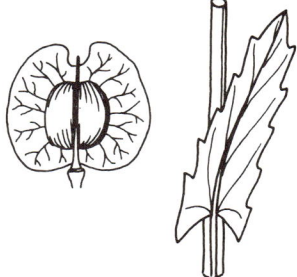

FIELD PENNY CRESS, pod and stem-leaf

The sight of FIELD PENNY CRESS, *Thlaspi arvense*, growing in your garden is enough to daunt the stoutest heart. There it stands, stems and stemlets laden with the round pennies of its capsules. The total number of seeds per plant is about 900. Console yourself, however, for the weed is an indicator of a good loam soil rich in nutrients, and it is not difficult to pull up. An annual or winter annual, it grows 1ft (30cm) tall and from a single stem it becomes much branched. All the leaves are alternate, the lowest ones being stalked and narrowly oval, soon drooping; the middle and upper leaves oblong, usually toothed, and clasping the stem. The small white flowers bloom from April to August. A native of Eurasia and the Mediterranean region, this troublesome weed is naturalized throughout the United States and north into Canada from Newfoundland to British Columbia and Alaska.

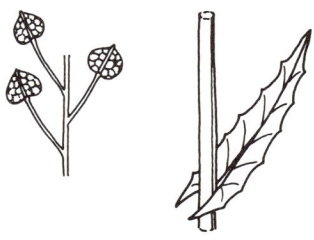

HOARY CRESS, pods and stem-leaf

In England the HOARY CRESS or PEPPER CRESS, *Cardaria draba*, is also known as THANET CRESS. It was introduced from Holland after the ill-fated Walcheren Expedition of 1809, when the fever-stricken soldiers were brought home on mattresses stuffed with hay which was bought by a farmer in Thanet, in east Kent, and ploughed in for manure. The Cress sprang up. Introduced into America the Hoary Cress spread throughout all the United States excepting the area along the southern boundary of the western and south-central States. Growing to a height of 1–3ft (30–90cm), though in America its height is less, the Hoary Cress reproduces not only by its seeds but by its numerous rhizomes sending up growing-shoots, these branching from the main root that can penetrate down to a depth of 10–12ft (3–3.5m) and develop new buds from any broken portion of root. Eradication is therefore difficult.

The Hoary Cress bears rather large flat-topped clusters of small creamy-white flowers from April to June, followed by broadly heart-shaped pods standing vertically on their stalks. The leaves, long and toothed, lobed at the base and clasping the stem, grow spirally around it. The stem is hairy, hence the plant's first common name. The Cress was formerly ground as a substitute for pepper, hence the synonym.

A much smaller weed is the annual Shepherd's Cress, *Teesdalia nudicaulis*, native to Britain and Europe and local in America, with flower-stems only 2–4in (5–10cm) high, growing from a rosette of pinnate leaves having 5 tiny leaflets. It may have an occasional pinnate leaf on the stem, which is otherwise leafless. Apart from its size, this Cress is recognizable by the pink sepals and the two outer white petals being longer. The seed pods are again heart-shaped. The flowering period is from April to June.

Shepherd's Cress, pods and rosette-leaf

An annual to biennial, Shepherd's Purse or Pickpocket, *Capsella bursa-pastoris*, has triangular heart-shaped seed capsules growing straight out from the stem, and the weed is taller, 12–16in (30–40cm), with its rosette leaves varying from pinnate to lanceolate and undivided. The alternate stem-leaves are also variable in shape, the undivided ones being lobed at the base and clasping the stem. The tiny white flowers are $\frac{1}{10}$in (2.5mm) across. Shepherd's Purse is a cosmopolitan from southern Europe, so common that almost every garden has it. In Britain it flowers all the year, in North America from March to December.

Shepherd's Purse or Pickpocket, pods and rosette-leaf

The majority of American weeds are introductions from Europe. The Virginia Pepper Weed, or Poor Man's Pepper, *Lepidium virginicum* is, as its name indicates, a native of North America which was introduced into Britain. It is an annual or biennial with a single erect stem 12–20in (30–50cm) tall, downy and with long downward-curving hairs. The stem is much branched at the top, and again the leaves are variable, the lower ones being pinnate with 4 or 5 pairs of toothed leaflets, the upper ones toothed but undivided, oval to lanceolate. It blooms in long racemes from May to November, the flowers being inconspicuous and white or greenish. The seed capsules are almost round.

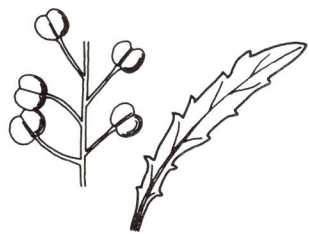

Virginia Pepper Weed or Poor Man's Pepper, pods and stem-leaf

Another name for Winter Cress, *Barbarea vulgaris*, is Yellow Rocket which aptly describes its dense inflorescence of bright yellow-petalled flowers shooting up the stem. It is a biennial or perennial with a thick, yellowish tap-root and erect, branching stem 1–3ft (30–90cm) tall. It has 3 kinds of leaves: pinnate ones with 5–9 narrowly oval leaflets growing at the base in a rosette; lower stem-leaves which have only a few lateral lobes; and, at the top, almost round crenate leaves which are undivided. All the leaves are deep

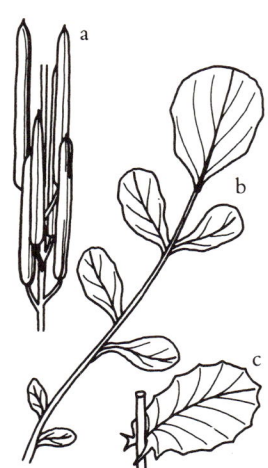

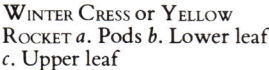

Winter Cress or Yellow
Rocket *a*. Pods *b*. Lower leaf
c. Upper leaf

green and shiny. The flower-spikes grow from the leaf-
joints, and the slender pods have light-yellow seeds. A
native of Britain and Europe, the Yellow Rocket grows
under this name throughout the north-eastern area of
the United States as far south as Arkansas, in distinct
areas in the north-central states, in Washington and
Oregon and north into Canada from Newfoundland to
Ontario.

What would the world be, once bereft
Of wet and wildness – let them be left,
O, let them be left, wildness and wet,
Long live the weeds and the wilderness yet !

Gerard Manley Hopkins *Inversnaid*

DAISY (Compositae)

One of the most familiar of all weeds in Britain — it is called the ENGLISH DAISY in North America, which tells its origin — the DAISY, *Bellis perennis*, is an infestor of lawns where its rosettes of leaves hug the ground and effectively smother the grass. These leaves are shiny, oval to spatulate, and round and broad at the end, narrowing to a short broad stalk at the base. The hairy flower-stalk is leafless and ends in a single rosy bud opening to a head of numerous white ray-florets surrounding a cushion of bright yellow disk-florets.

The common name comes from Day's Eye, as it is also called, for the little flower, $\frac{4}{5}$–1in (16–25mm) across, closes at sundown. Bairnwort is another name for it, and children still love to make daisy-chains by slitting the stems and threading the flowers through them, while Bruisewort was the name given to it in the Middle Ages by the Crusaders who used it for healing their wounds and bruises. Many are the legends and superstitions connected with the Daisy, and it is the herald of spring — for spring, it is said, has not arrived until you can put your foot on twelve of these flowers, which goes to show how they can colonize your lawn! Pest as it is, the Daisy is loved too; but one can have too much of a good thing, and a selective weed-killer will successfully remove it.

DAISY or ENGLISH DAISY

DANDELION (Compositae)

The DANDELION, *Taraxacum officinale*, is almost as common as the Daisy, and much more troublesome — unless you can make use of it as a culinary or medicinal herb: it contains more Vitamin C and A than almost any other vegetable or fruit, and its bitterish leaves, shredded or chopped roots and sweet tangy flowers add value to any salad. Entirely beneficial to man, stimulating the bloodstream, liver, digestive organs and especially the kidneys and bladder, the Dandelion won fame as a 'potty herb': Pee-a-bed, Wet-a-bed and, in France, Pissenlit were names given in this connection. As a garden weed it is not to be encouraged, for it absorbs about three times as much iron from the soil as is taken up by any other plant, with copper and other soil nutrients. Hence its value in a salad. But it makes a bad neighbour for other plants also by exhaling ethylene gas which hinders their growth.

We can, however, turn its thefts to good account by composting the plant. Rotted down, the stolen riches can be fed back into the soil where shallower-rooted garden plants can absorb them. It can also be made into a liquid fertiliser and into a folia-feed to make good the deficiences in garden plants and at the same time act as an insect-repellent.

The Dandelion's tooth-like lobed leaves are confined to a basal rosette growing from a simple or branched tap-root. The flowers, in bloom most of the year, are heads of yellow florets, solitary on a hollow, sometimes red stem containing a white milky juice. They are visited by a great variety of insects, and the flower is followed by 'the globe of down, the schoolboy's clock in every town' which at a puff separates into scores of seed-carrying parachutes. More accurately the Dandelion tells the time by opening its flower between six and seven in the morning. A native of Eurasia, it is a cosmopolitan weed abundant throughout the northern hemisphere.

DANDELION or COMMON DANDELION

DOCKS (Polygonaceae)

The great American botanist Asa Gray described Docks as 'coarse herbs, with small and homely (mostly green) flowers' crowded and commonly whorled around the stalks in simple or branched inflorescences. Natives of Europe, they have spread to most parts of the world, and they are masters of survival. They are perennials, and their seeds (up to 50,000 of them annually per plant) are wind-borne, and can also be carried by birds, and even underground by ants for use as nesting material, these buried seeds remaining viable for as long as 80 years. The branches of the thick yellow tap-roots may reach down 5ft (1.5m) or further, and any broken portion will readily shoot and become a new plant.

Five species plague gardens. Though much alike they can be distinguished by their leaves.

The CURLED or CRISPED DOCK, *Rumex crispus*, known in the United States as the YELLOW or CURLY DOCK, is the commonest and most troublesome. It has waved and twisted leaves, broadly lanceolate, up to 1ft (30cm) long, arranged alternately up the stem. Its height ranges from 20 to 42in (50-100cm); the flowering period is from June to October. The branched flower-spikes bear the flowers in whorls intermingled with few-to-many ribbon-like leaves, again wavy. A large plant can produce 30,000 seeds in a year, of which 88 per cent will germinate. A 52 per cent germination has been recorded with seeds buried in the soil for 50 years. This is a notifiable weed agriculturally in Britain, farmers being required to eradicate it. Its presence in your garden tells that the soil is rich in nutrients.

The Curled Dock is an old inhabitant in Britain: it has been found in deposits of the Palaeolithic Age, the earliest known culture period in Europe, whence in more modern times it was introduced into America.

CURLED DOCK or YELLOW DOCK, inflorescence and leaf

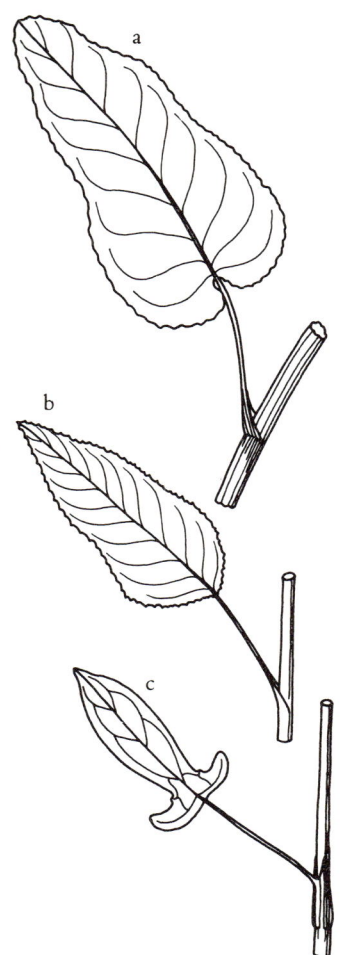

Leaf of *a*. BROAD-LEAVED or BITTER DOCK *b*. SHARP DOCK or SMALLER GREEN DOCK *c*. SHEEP'S SORREL or RED SORREL

The BROAD-LEAVED DOCK, *Rumex obtusifolius*, known also in North America as the BITTER DOCK, is almost as common in Britain and ranges in western Europe from northern Spain to southern Scandinavia, and eastwards to central Germany and Hungary. It has alternate leaves about 10in (25cm) long, broadly oval with a heart-shaped base. Again the inflorescence is in whorls around the stems which are branched and rather spreading, leafy in the lower part. The perianth surrounding the fruit has 3 to 5 long teeth along the margin each side. The flowering period is from June to October. This species is also a European introduction into the United States.

The SHARP DOCK, *Rumex conglomeratus*, the SMALLER GREEN DOCK of the United States, has long oval leaves which, though narrower, resemble those of the Curled Dock. Distinguish this one then by the flower-clusters growing close to the stalks and stem, and the winged appearance of the fruit. The plant grows up to 39in (100cm) tall. A native of Europe, it is naturalized in the United States.

Rumex acetosella is SHEEP'S SORREL in Britain and North America where because of its red stem and often red leaves it is also called RED SORREL. It too has a tough tap-root, and the horizontal roots spreading from it just below the surface are a source of shoots which rapidly become colonies of new plants. Its smaller height, up to 1ft (30cm), and its arrowhead alternate leaves, acid-tasting, distinguish it easily from other docks, as do the flowers which are yellow to red, borne loosely on erect stems from May to August. Sheep's Sorrel is an indicator of an acid soil of poor fertility.

Dull and homely docks may be, but in summer when the seeds are ripe the Sheep's Sorrel blazons with rust-red spikes and scarlet leaves.

The Sorrel, *Rumex acetosa*, the Garden Sorrel of America, has smooth, oval, alternate leaves with two incurving lobes at the base. Bitter-tasting, as its Canadian name Sour Dock tells us, its leaves are sometimes used in salads and sauces. The flowers, in May and June, are followed by fruits that are peculiarly shining. Up to 39in (100cm) tall, the Sorrel is generally distributed and common throughout Britain, and ranging through Europe, temperate Asia, and North America into Greenland.

Leaf of Sorrel or Garden Sorrel

Enchanter's Nightshade (Onagraceae)

The Common Enchanter's Nightshade, *Circaea lutetiana*, is an odd-looking plant, with its small white flowers hanging from long stalks arranged alternately on a spike. These develop into pear-shaped fruits densely covered with glistening hooked bristles that readily catch on to clothing or the fur of small animals. The flowers themselves are curious, for they have only 2 petals, though these are so deeply lobed as to look like four. From the slender root-stock the stem, which is erect and has pairs of pointed oval leaves, grows to a height of 8–28in (20–70cm). This is a shade-loving plant flowering in July and found throughout Great Britain and Europe.

The generic name comes from Circe, the Greek enchantress who turned the followers of Odysseus into swine by giving them a magic drink, but despite its name the Common Enchanter's Nightshade is unrelated to the Deadly Nightshade and is not poisonous.

Common Enchanter's Nightshade

FLEABANES (Compositae)

The travels of the HORSEWEED or CANADIAN FLEABANE, *Conyza canadensis*, in America *Erigeron canadensis*, make fascinating history. A native of North America, it was brought to Europe about 1686 in the skin of a bird stuffed with its plumed seeds. It was recorded as growing in the London area some 20 years later, and by 1877 was common on waste grounds in London and on the embankments of the new railways built in the early forties, carrying traffic to and from the capital. By the end of the century it was growing in ten counties south of the line from the Bristol Channel to the Wash. During World War II it was a familiar feature of city bombed sites.

A stiff erect annual, 1–2ft (30–60cm) high, it has many branches of small tubular disk-florets fringed with minute ray-florets, the outer ones threadlike and greenish-white to lavender or slightly tinged with red, the centre ones yellowish-white. The narrow dark-green leaves grow alternately up the stem, and both stem and leaves have bristly hairs. It flowers from July to November, after flowering transforming itself into a mound of white fluff. A prodigious seeder, large plants can produce more than a quarter of a million fruits. Treat it with some respect, for it contains a terpene that irritates the skin, and it is suspected of poisonous properties if eaten. As a contribution to the compost heap it adds some nitrogen.

PHILADELPHIA FLEABANE, *Erigeron philadelphicus*, is similarly spread throughout North America but has not so far found its way to Britain. It is a short-lived perennial with stems 2–4ft (60–120cm) tall. The stems, which are hairy, bear branches of small heads of flowers 1in (2.5cm) across, having many very narrow rose-purple or flesh-coloured ray-florets, the disk-florets being yellow. The numerous leaves, which are alternate, are lanceolate with a broad midrib and scattered with coarse white bristles. The Philadelphia Fleabane likes moist ground, flowers from June to November, and produces seeds with bristles at one end.

a. HORSEWEED or CANADIAN FLEABANE *b*. Leaf of PHILADELPHIA FLEABANE

FORGET-ME-NOT
(Boraginaceae)

The COMMON FORGET-ME-NOT, *Myosotis arvensis*, is not a very troublesome weed though it tends to steal into the garden and spread rapidly, each plant yielding about 700 seeds and having two flowering periods, May to July and August to October. But it is easily uprooted. It is recognizable by its bright blue flowers, $\frac{1}{5}$in (5mm) across with 5 petals and a yellow eye. These grow alternately on short stalks up the long hairy stem. Before flowering, the stems are curved and rolled inwards at the ends. The lower leaves are stalked, roundish-oval, and form a rosette at the base; the stem-leaves grow alternately and are lanceolate, clasping the stem. All the leaves are hairy on both surfaces. The Common Forget-Me-Not is an erect annual 6–12in (15–30cm) in height. It is common over practically the whole of Europe, northern and central Asia and North America. A loam-indicator, it likes a well aerated soil, more or less rich in nitrogen.

COMMON FORGET-ME-NOT

FUMITORY   (Fumariaceae)

The name comes from *fumus*, smoke, the ancient exorcists believing that the smoke from fumitories had the power of expelling evil spirits. The COMMON FUMITORY or EARTH-SMOKE, *Fumaria officinalis*, is a dainty little weed with spikes of 10–40, usually more than 20, pink tubular flowers, darker pink near the tip, growing alternately on long stalks from the slender stem. The soft grey-green leaves, growing in an almost whorled arrangement on the same stem, are very much divided into linear leaflets. The whole plant has a delicate feathery appearance and flowers nearly all the year round, each plant yielding about 800 seeds. A native of Europe and western Siberia it has been introduced into America. It is an annual growing up to 12in (30cm) high and likes a loose, nutrient-rich and usually lime-deficient, loamy soil.

COMMON FUMITORY or EARTH-SMOKE, with detail of inflorescence

GALLANT SOLDIER
(Compositae)

Introduced into the United States from Mexico, the SMALLFLOWER GALINSOGA took on a more jaunty name after it was found growing in a garden at Richmond, England, in 1860. It was an escape from nearby Kew Gardens where it had arrived in a consignment of new plants. The name was GALLANT SOLDIER from its Latin designation *Galinsoga parviflora*. It also became known as Joey Hooker, he being the son of Kew's director Sir William Hooker, so presumably it was Joseph who recorded its appearance as a weed. The seeds are borne on parachutes of silvery scales and dispersed by the wind but easily adhere to clothes.

The Smallflower Galinsoga is an annual with a much-branched erect stem 4–30in (10–75cm) tall. The pointed oval leaves grow in pairs, and from their axils spring the stalked flowers which look like tiny marigolds about $\frac{1}{4}$in (6mm) across with 4–6 ray-florets, most frequently 5, white and 3-lobed at the tip, surrounding 10–60 yellow tubular disk flowers.

In flower from May to November it is a rapid grower, and each plant produces between 5,000 and 30,000 seeds which can be in flower 4 weeks after germinating. No wonder this weed is now distributed over almost the entire world. It occurs on lime-containing loams but prefers a lime-deficient, neutral loamy soil rich in nutrients.

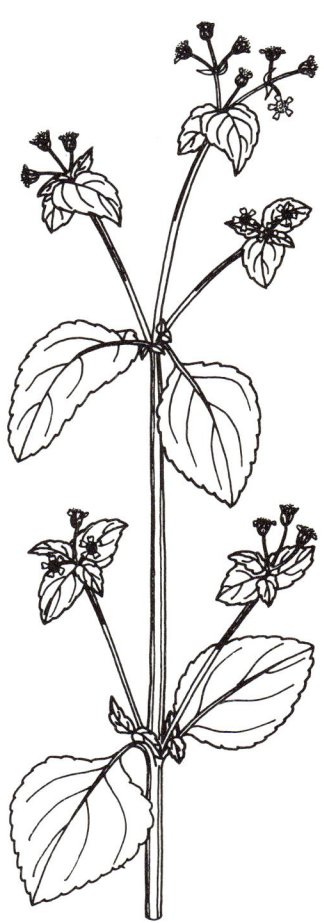

GALLANT SOLDIER or
SMALLFLOWER GALINSOGA

GARLIC (Liliaceae)

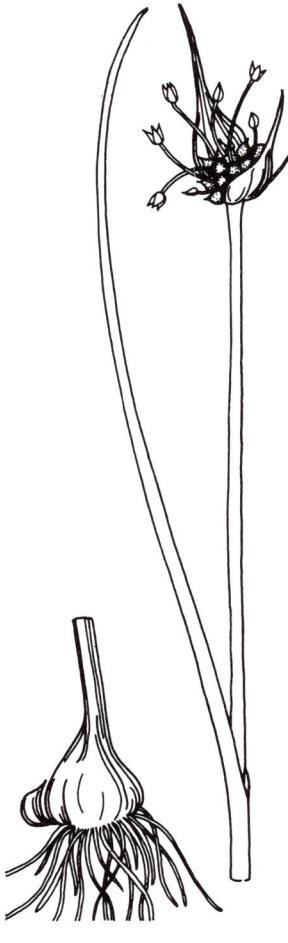

A terrible weed is *Allium vineale*, the CROW GARLIC, or FIELD GARLIC as it is called in the United States and Canada. It reproduces in five ways: by seeds in the spring which germinate in the following autumn, and by four kinds of bulbs which are found at the end of the season in late spring — aerial bulbils growing in a cluster at the top of the long thin stem; hard-shell bulbs formed underground in the axils of the outer leaves; central bulbs, also formed underground but around the main axis of the plant; and soft offset bulbs, the largest of the four types, formed around the parent bulb. It does not, however, produce all these at once: there are two types of the plant, one bearing a cluster of aerial bulbils and 6-petalled flowers sometimes, greenish-white, pink or purplish followed by black seeds; the other a smaller plant with bulbils but without flowers. Rarely does the plant produce flowers only. The long hollow leaves have sheathing bases and are striped. The height varies from about 1 to 3ft (30–90cm).

With all its means of survival the Crow Garlic is difficult to eradicate, especially when it infests lawns, and it is drought-hardy and cold-hardy. It likes a heavy soil and is tolerant of poor drainage. A native of Eurasia (Europe and Asia) it is common in England and Wales and is found throughout most of the eastern and central states of the United States, along the Pacific Coast from Washington to northern California, with a distinct area in Wyoming. It is common in the eastern part of Canada.

CROW OR FIELD GARLIC

GERANIUMS (Geraniaceae)

A delightful family are the Geraniums, comprising that charmer of pink leaves and flowers, Herb Robert, and the cranesbills and storksbills — so-named from the long beak of the ovary, visible when the petals have fallen. The storksbills (*Erodium*) are distinguished from the cranesbills (*Geranium*) by having pinnate, more feathery leaves, umbels instead of single or twin flowers, and much longer beaks. The wild Geraniums are not to be confused with the garden and greenhouse ivy-leaved 'geraniums' which belong to the genus *Pelargonium* and should commonly be so called.

The lower leaves of the CUT-LEAVED CRANESBILL, *Geranium dissectum*, grow on long stalks in pairs and are deeply divided into 5–9 lobes. These lobes are in turn lobed as single, bifid or trifid leaflets. The 5 reddish-pink petals are notched, looking twice the number. An annual, the plant is branched, the stems densely hairy, and often of a straggling habit. It flowers from May to August and its presence indicates a nutrient-rich soil. A native of Europe, it is common throughout the continent and the British Isles and is naturalized in North and South America.

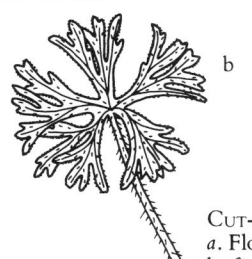

CUT-LEAVED CRANESBILL
a. Flowering stem *b*. Lower leaf

All the cranesbills are relatively resistant to selective herbicides, although some control can be achieved by spraying in the early seedling stage if they become a nuisance. Otherwise they are easily uprooted by hand.

The CAROLINA CRANESBILL, *Geranium carolinianum*, has hairy stems branching from the base and small, pale, rosy flowers with 5 petals, not notched, that bloom

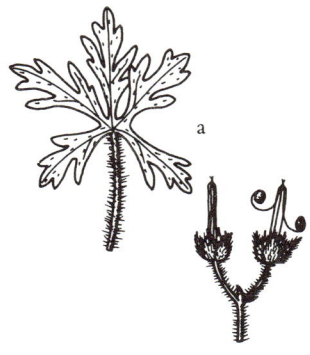

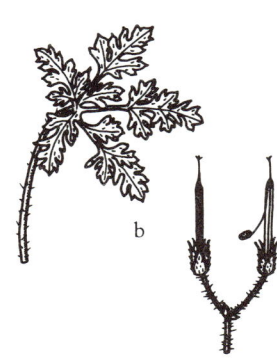

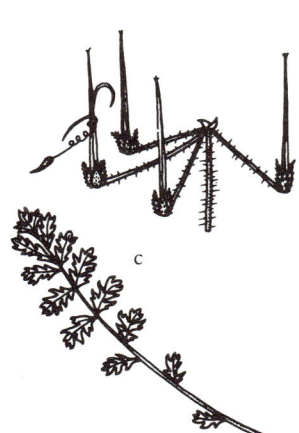

Leaf and fruit of *a*. CAROLINA
CRANESBILL *b*. HERB ROBERT
c. COMMON STORKSBILL or
ALFILERIA

from May to July. The leaves, growing alternately up the stem, have 5 divisions cleft and cut into numerous oval-linear lobes. The stalks of the clustered inflorescence and flowers are short. If you have this weed in your garden it will tell you that the soil lacks nutrients, for the Carolina Cranesbill can tolerate even barren soil. As its name indicates, it is a native of the South and South-east Atlantic region of the United States whence it has spread to every state as the worst weed of all the Geraniums that invade gardens.

HERB ROBERT, *Geranium robertianum*, is the pretty little plant recognizable by its red stems, which are hairy and succulent, much-divided leaves that towards autumn turn the same colour, and pink flowers with 5 petals that are undivided, the whole plant having a strong foxy smell. It is an annual or biennial, and its stems, branching from the base, grow 4–20in (10–50cm) high. The opposite leaves have 3–5 lobes, crenate round the margins. We do not know for certain how Herb Robert got its name, whether from Robin Hood (which is one of the plant's other names) or from St Robert of Molesmes who is supposed to have cured a certain disease with it. This disease, once called Ruprecht's Plague in Germany, gave rise to the plant becoming known there as Ruprechts-Kraut. But, again in Germany, it is also called Roberts-Kraut. Herb Robert is a European that has become naturalized in North and South America.

The COMMON STORKSBILL, in America ALFILERIA, *Erodium cicutarium*, is a straggling plant with prostrate hairy stems, having much the habit of the Geraniums but distinguished immediately by its feathery pinnate leaves and umbels of lilac or rosy-purple flowers sometimes with a blackish spot at the base of the two upper petals. In fruit the beak measures $\frac{7}{8}$–$1\frac{5}{8}$in (22–40mm). An annual, occasionally biennial, the Common Storksbill varies in height from 3in to 2ft (7–60cm), its pinnate leaves varying in length from under 1in to over 7in (2–20cm). These are opposite, the leaflets on each leaf being alternate and themselves pinnate. The plant is in flower all the summer and is widespread in Great Britain and Europe, naturalized in North and South America. It is an indicator of sand and nitrogen.

GOOSEFOOTS
(Chenopodiaceae)

The Goosefoot family is recognizable by the shape of the leaves, the botanical name coming from the Greek *chēn*, a goose; *pŏdion*, a little foot. The markings on the testa or outer skin of the seeds provide features distinguishing between the different species, and these can be seen with a magnifying glass. But so far as the gardener is concerned, the differences of leaf and inflorescence are sufficient for identification. Many of the goosefoots are edible: the red garden beetroot, spinach, mangolds and sugar beet belong to the family. Fat Hen makes nutritious eating cooked as spinach, and until the introduction of this vegetable Fat Hen was the most valued for human beings and for animal fodder.

FAT HEN, *Chenopodium album*, called COMMON LAMB'S QUARTERS in the United States and Canada, is an annual varying greatly in height: it may be only a few inches tall or up to 6ft (nearly 2m). The stems, often reddish, are simple or branching with green or purplish ridges. The leaves, which are alternate, vary in shape, the lower ones being the characteristic goose-foot but not lobed; the upper ones sometimes linear, growing close to the stem and usually mealy-coated, especially on the underside and in the young stage. A first glance at this weed is almost enough to identify it: commonly it is a much-branched pyramidal plant crowded with clustered spikes of dull green flowers. The testa of the seed has either very faint branching ridges, being otherwise nearly smooth, or sometimes has raised lines forming irregular cell-like markings.

The flowering period is from June to October and each plant produces about 3,000 seeds but can produce up to 20,000. It is by far the commonest Goosefoot throughout the British Isles, the United States and Canada. Its distribution is world-wide and it thrives on loose, damp, nitrogenous loams or sandy soils and extracts large quantities of mineral nutrients. These it raises in solution from deep in the ground, storing them in its roots and leaves. By composting Fat Hen the nutrients can be returned to the soil, but this time within reach of shallower-rooted garden plants.

a

FAT HEN or LAMB'S QUARTERS
a. Surface of seed testa

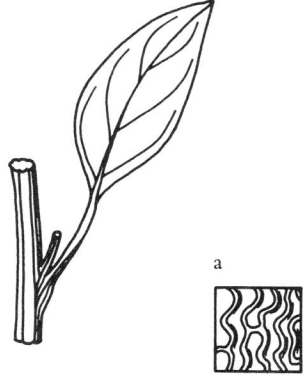

Leaf of ALL-SEED or
MANY-SEEDED GOOSEFOOT
a. Surface of seed testa

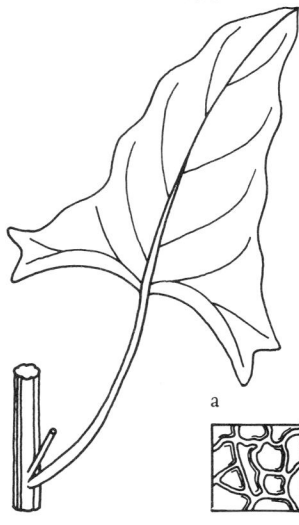

Leaf of GOOD KING HENRY
a. Surface of seed testa

ALL-SEED or MANY-SEEDED GOOSEFOOT, *Chenopodium polyspermum*, is an annual with stems erect or decumbent, red and usually 4–angled. It can attain a height of up to 40in (about 1m). The leaves, about 2in (5cm) long, are oval or elliptical and grow alternately up the stems. They may be undivided in this species or with a single tooth-like angle on one or both sides towards the base. The tiny green flowers are arranged in clusters on slender spikes growing from the leaf-axils. The testa of the seed has pits with wavy margins. Each plant produces about 4,000 seeds following the flowering period of July to September. All-seed likes a well-aerated moist and friable soil, weakly acid to alkaline and rich in nitrogen. It is a native of Eurasia occurring locally in North America.

GOOD KING HENRY, *Chenopodium bonus-henricus*, has its lower leaves of the characteristic triangular shape, with lobes making acute angles with their stalks, the leaves above being narrower and making obtuse angles. The tiny green flowers are in clusters on tapering spikes which are leafless except at the base. All the leaves are alternate, thickish and of a dark green except in the young stage when they are mealy. The plant is an erect perennial growing to a height of 12–20in (30–50cm). The testa has a rough, stippled appearance. Good King Henry, whose other names are All-Good and Mercury, is listed as an introduction into Britain, but this was a very long time ago: seeds are found in Bronze Age and Roman deposits. It was cultivated for eating as a vegetable, and as a healing herb. In Europe it grows as far north as central Scandinavia and east as far as western Asia. It has crossed the Atlantic to North America but is not common. The flowering period is from May to August, and it likes a nitrogen-rich soil.

IRON ROOT or COMMON ORACHE, *Atriplex patula*, is recognizable by the stems often being striped white and green or red and green, and by the lower leaves being the true Goosefoot shape: that is, 3-lobed, the middle lobe longer and with toothed margins, the two lower shorter lobes making obtuse angles with the leaf-stalk. The upper leaves are not lobed and are narrowed and smaller as they grow up the stem. The two genera *Chenopodium* and *Atriplex* differ in their flowers: the first being a hermaphrodite, with stamens and ovaries in the same flower; while *Atriplex* has separate male and female flowers, though on the same plant. The Orache is an annual of sturdy growth, sometimes erect and ascending but more often sprawling. The stems are much branched and up to 32in (80cm) long. The green flowers form small clusters on long spikes which grow from the leaf-axils. It flowers from July to September and is distributed throughout the British Isles and Europe, the United States and Canada.

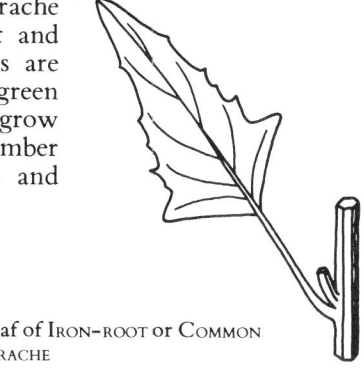

Leaf of IRON-ROOT or COMMON ORACHE

He that can draw a charm
From rocks, or woods, or weeds,
or things that seem
All mute, and does it − is wise.

Bryan Waller Procter
A Haunted Stream

GRASSES (Gramineae)

'How *green* ... !' Thus visitors to Britain are sure to exclaim on first seeing the lush meadows, pastures and even hillsides of these islands; and certainly British people themselves are impressed by the vivid emerald that spreads over the countryside in spring, first in the fields of winter wheat which are unbelievably brightly green, and then along roadsides, hedge-bottoms, waste places, pond-margins and indeed anywhere that a root can find a toehold.

Grass is responsible for much of the green, but gardeners on seeing their gravel paths green-filmed with young blades of it are not so enthusiastic. For grasses are well-known as terrible seeders. They need no butterflies or bees or other insects to pollinate them. This they do themselves, aided by the wind; and the moment the seeds are ripe, small birds like sparrows and finches greedily eat them and scatter them far and wide. It is no comfort to learn that throughout the world there are about 620 genera and 10,000 species of grasses. On the other hand we must tell ourselves that we rely on grasses for much of our staple food: wheat, oats, rye and rice all belong to the great family of Gramineae. Nor could the animals live without grass. It behoves us to know at least which grasses are most troublesome to us as weeds and what use we can make of them. Even Couch Grass, in the United States called Quack Grass, can be a valuable ingredient of the compost heap.

ANNUAL POA or ANNUAL MEADOW GRASS, *Poa annua*, in North America LOW SPEAR-GRASS or ANNUAL BLUE-GRASS, varies in size and height according to its environment, from 2 to 12in (5–30cm) tall. The leaf-blades are strap-shaped, meaning that they are practically the same width for their entire length, and terminate in a boat-shaped tip. The panicle or branched flower-head is triangular in outline with small spikelets of 3–10 flowers, and seeds are shed throughout the year, even occasionally in winter. The first leaf is erect, with a broad upper part narrowing abruptly to a blunt point. Characteristics are the two parallel lines on the blade close to the central fold of the leaf: these are known as

ANNUAL POA or LOW SPEAR GRASS

'tramlines' and they can clearly be seen when the leaf is held up to the light; also, the leaf-blades are often wrinkled or puckered. These with the boat-shaped leaf-tip identify this grass.

A native of Eurasia, *Poa annua* is distributed throughout nearly the whole world, even in the tropics where it grows mainly on mountains; this is the common little grass that grows in town as well as in the country, in the cracks in pavements and on garden paths and in lawns, although it is not an ideal lawn grass because of its soft blades and often lax habit.

MEADOW GRASS, *Poa pratensis*, is the famous KENTUCKY BLUE-GRASS on which American thoroughbreds graze. It is a rather stiff perennial, tufted, erect, and varying in height from 4 to 36in (10–90cm). It has slender creeping rhizomes, and the leaves are green or greyish-green, smooth or sometimes rough and usually with a tip abruptly contracted. The panicle is composed of thread-like branches bearing several short branchlets which carry the egg-shaped spikelets of a pale greyish-green streaked with dark green and yellow. It is in flower from May to July. *Poa pratensis* occurs commonly throughout the British Isles; in continental Europe, growing only on mountains in the south; in temperate Asia, North Africa and North America.

MEADOW GRASS or KENTUCKY BLUE-GRASS

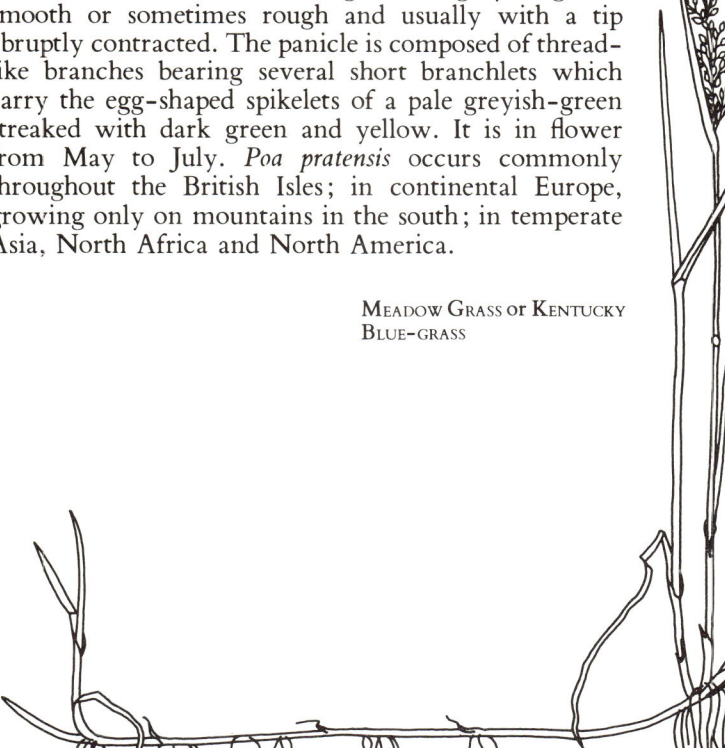

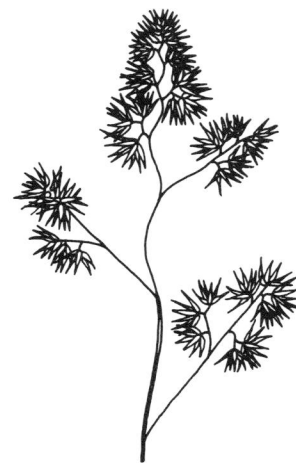

Panicle of Cock's Foot or
Orchard Grass

COCK'S-FOOT, *Dactylis glomerata*, called ORCHARD GRASS in North America, is coarse-looking and tufted, often bluish-green in colour and varying in height from 1 to 4ft (30cm to just over 1m). It is easily recognized because the panicles first appear in May and are in flower all summer until October. The panicle has a characteristic feature, the spikelets being crowded together in several dense clusters. The blades are long and flat, narrowing to a point and rough on both sides with fine teeth on the edges. In the seedling stage the top of the first leaf bends over and the leaf-tip narrows to a blunt point. An important pasture and hay grass, the Cock's-foot or Orchard Grass was introduced from Europe into Britain and is distributed throughout mid-Canada and the southern states. It is not so welcome in gardens, though the variety *variegata* with striped green and white leaf-blades is sometimes cultivated.

COUCH GRASS, known on both sides of the Atlantic by this name, is one of the real villains in a garden. Witch Grass, Twitch and Quack Grass are other names for it. By whatever name, *Agropyron repens* is noxious and persistent, a perennial spreading not only by seeds but by the underground rootstock, almost any part of which is able to produce a new plant. Sooner or later every gardener meets the abundant and far-creeping white or yellowish rhizomes winding their way through the soil, under and over and through the roots of treasured plants, and seemingly endless. Know it above-ground by its dull green blades, long and tapering to an acute point, growing alternately up the stiff stem, and by the flower-head with its spikelets arranged alternately and close to the stalk. A native of Europe, it is common throughout the British Isles, Canada, and throughout all the United States excepting the south.

COUCH GRASS

A serious weed in lawns and garden borders is the HAIRY FINGER GRASS, *Digitaria sanguinalis*, called LARGE CRAB GRASS in the United States and HAIRY CRAB GRASS in Canada. It was introduced into Britain from southern Europe as an impurity in grain, in the same way spreading throughout South America and the United States excepting an area between Lake Superior and the Rockies. The spikelets radiate from a single axis like the fingers of a hand, hence the generic name for this grass, and one of its common names. It is a loose-growing green or purplish annual 4–20in (10–50cm) tall, at first erect but soon becoming decumbent and spreading by rooting at the nodes of the culms (stems) lying along the ground. The leaves, which are narrowly lanceolate, rounded at the base and finely pointed, are flat and the lower ones hairy, as are the lower parts of the branching stems. It flowers from August to October.

The common names for *Digitaria ischaemum*, SMOOTH FINGER GRASS and SMOOTH CRAB GRASS, describe a distinguishing feature: neither stem nor leaves are hairy. This species also has distinctive inflorescences: whereas the spikelets of the hairy species arise from a single point, those of *ischaemum* come in pairs or threes not all arising from the same point. The grass is also known as Red Millet. A native of Eurasia, it is rare in England but found throughout all the United States excepting southern Florida and parts of the south-west along the Mexican border. It flowers from July to October, and is another grass weed troublesome in lawns.

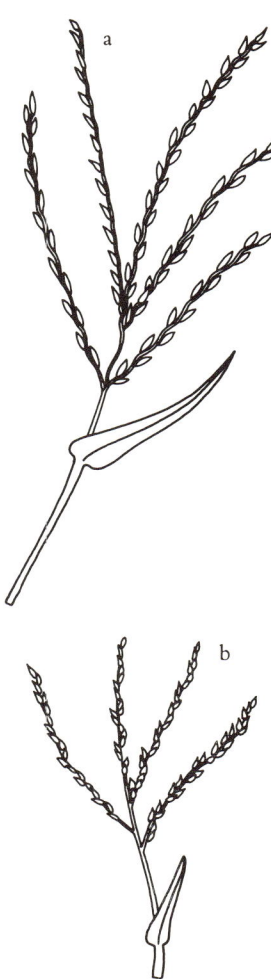

Panicle of *a*. HAIRY FINGER GRASS or LARGE CRAB GRASS *b*. SMOOTH FINGER GRASS or SMOOTH CRAB GRASS

CREEPING SOFT GRASS, *Holcus mollis*, and MEADOW SOFT GRASS, *Holcus lanatus* (better known in Britain as YORKSHIRE FOG), are so similar that it is better to describe them together. They are given the name Soft Grass because of their woolly appearance: in the United States and Canada *Holcus lanatus* is called VELVET GRASS; and in the United States *Holcus mollis* is the GERMAN VELVET GRASS, recently introduced from Europe but spreading. Both are perennials and both range in height from 8 to 40in (20cm to just over 1m); both have the same sharply pointed blades 1½–8in (4–20cm) long, and oval panicles.

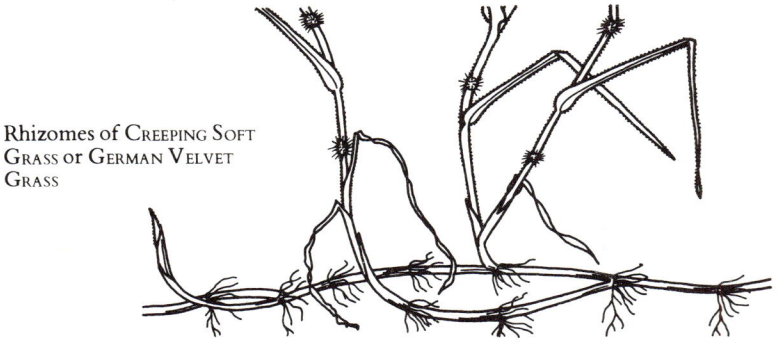

Rhizomes of CREEPING SOFT GRASS or GERMAN VELVET GRASS

Here are the differences between them: *Holcus mollis* — extensively creeping rhizomes; bearded nodes; panicles whitish, pale grey or purplish: *Holcus lanatus* — fibrous-rooted; nodes not bearded; panicles whitish, pale green, pinkish, or purple. *Holcus mollis* flowers from June to August, *Holcus lanatus* from May to August.

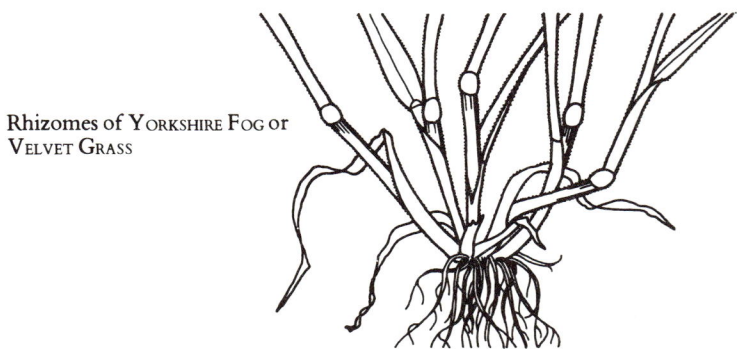

Rhizomes of YORKSHIRE FOG or VELVET GRASS

FALL PANICUM, *Panicum dichotomiflorum*, is an American grass flourishing throughout all the United States excepting parts of the north-central and north-western states and Texas. It is an annual flowering from June to October, 20–40in (50cm to just over 1m) high but occasionally taller. The blades are narrowly lanceolate and rough to the touch, 4–20in (10–50cm) long, with a prominent white midrib. The panicles of alternately branched spikelets are loose in habit, but the main branch ascending. The epithet *dichotomiflorum* means that the florets are forked in pairs, and this is a distinguishing feature. An annual, the Fall Panicum is fibrous-rooted.

OLD WITCH GRASS, *Panicum capillare*, grows up to about 2ft (60cm) tall and is easily recognizable by its panicle which emerges from the wide sheath like a pointed shaving-brush and opens into a wide-spreading, much-branched inflorescence with florets on thread-like stalks, each carrying a single seed. The leaves are about $\frac{1}{2}$in (1.5cm) wide, broad at the base, hairy and tapering to the tips. The panicums are mainly tropical or sub-tropical grasses, but Old Witch Grass sometimes occurs abundantly as a weed in carrot fields in East Anglia. In eastern North America it is regarded as a native.

Another American grass is *Eleusine indica*, variously called GOOSEGRASS, WIREGRASS, YARD GRASS and SILVER CRAB GRASS. Though naturalized from the Old World it does not occur in Britain: its homeland is much farther south in a warmer climate. In appearance like the Finger Grasses or Crab Grasses, the spikelets of the Goosegrass are denser; it is fibrous-rooted and does not spread by rooting at the nodes, reproducing itself only by seeds. Also it is darker green and grows in tufts. An annual, flowering from July to October, it is a common weed in lawns and gardens throughout all the United States excepting northern Maine and parts of the north-central and north-western areas.

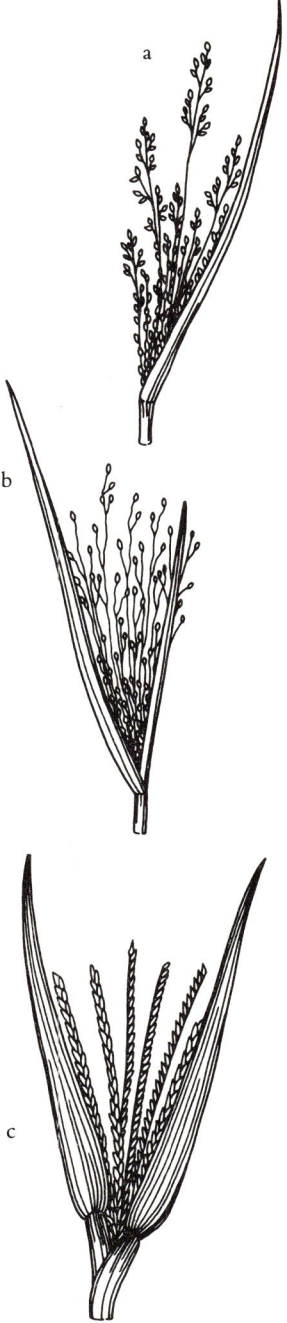

Panicle of *a*. FALL PANICUM
b. OLD WITCH GRASS
c. GOOSEGRASS or WIRE GRASS

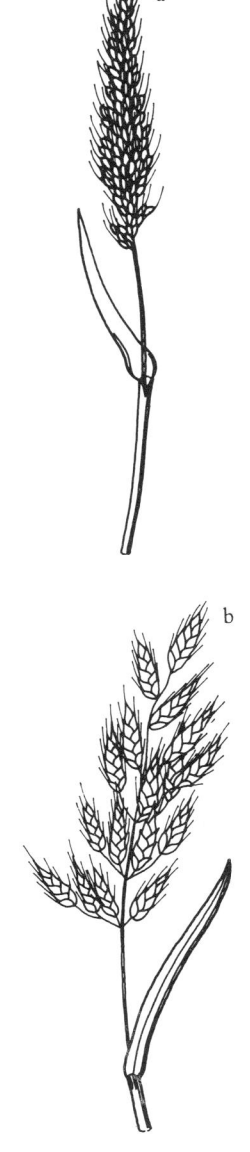

The GREEN BRISTLE GRASS, *Setaria viridis*, known in the United States by this name but more popularly as the GREEN FOXTAIL or BOTTLE GRASS, is one of the most serious and widespread grass weeds of cultivated soil in North America. The spikelets, which are green to purplish, are crowded together on a tail-like spike, and beneath each flower are protruding bristles which give the inflorescence its foxtail appearance. An erect annual, more or less tufted, this grass varies in height between 6 and 16in (15–40cm). Branching at the base and sometimes spreading, the Green Foxtail has flat blades usually less than 6in (15cm) long emerging from sheaths arranged alternately up the stems. It is fibrous-rooted. An introduction from Europe, it is common throughout all the cooler parts of the United States and Canada. In Britain it is a casual intruder near ports. The flowering period is from June to September.

LOP GRASS or SOFT BROME, *Bromus mollis*, is the most widespread of all the British species of this genus and is also found throughout continental Europe. It has been introduced into North and South America, where it is known as SOFT CHESS. It is an annual or biennial and its height ranges greatly, between 2 and 32in (5–80cm). The panicles have a wheat-like appearance, the flowers overlapping each other, but the spikelets are not as dense, nor the spike so stiff. Each flower has a projecting bristle. The blades are greyish-green, flat and limp, long and finely pointed with short soft hairs. Flowering from May to July, the Lop Grass begins to shed its seed early in summer.

Panicle of *a*. GREEN BRISTLE GRASS or GREEN FOXTAIL *b*. LOP GRASS or SOFT CHESS

NIMBLE WILL, *Muhlenbergia schreberi*, is another troublesome weed in American lawns, particularly in the eastern and central areas of the United States, central Colorado, and Mexico. Its slender stems are diffuse, branching, spreading and decumbent at the base, often rooting along the lower nodes. The blades are flat, usually less than 2in (5cm) long, and less than $\frac{1}{8}$in (3mm) wide. The panicles are long and thread-like, the alternate arrangement of the flowers making a slight zigzag. A perennial, the whole plant has a delicate, attenuated, lax appearance — this in the autumn, for in spring and early summer the stems are short and erect with spreading blades, making the grass look very different. It is not known in Britain or continental Europe.

Belonging to the Cyperaceae, the Sedge family, the YELLOW NUT SEDGE, or YELLOW NUT GRASS, *Cyperus esculentus*, is native to North America and spread throughout most of the United States; north to Nova Scotia, southern Quebec, southern Ontario, and southern Manitoba; south to Mexico and tropical America. It is unknown in Britain but is found in continental Europe and Asia. A serious weed in gardens, it likes a rich or sandy soil and reproduces both by seeds and by thread-like stolons ending in nut-like tubers. The triangular stem is erect and 8–36in (20–90cm) tall. From it fountain the long pale-green leaves. The flower-spike rises from the middle, a straight stiff stalk bearing the inflorescence in a terminal umbel like an inside-out umbrella, the yellowish to golden-brown spikelets being borne at the ends of long stalks in opposite pairs. Also growing from the axis of the umbel, and much exceeding the length of the flower-stalks, are long stiff leaves. The seed capsules or achenes are yellowish-brown, 3-angled and lustrous. A perennial, the Yellow Nut Sedge flowers from July to September. Although it has never naturalized in Britain, it was imported from the Mediterranean in medieval times as a favourite spice, hence the epithet *esculentus*, meaning 'edible'.

Panicle of *a*. NIMBLE WILL
b. YELLOW NUT SEDGE or YELLOW NUT GRASS

GROUND ELDER
(Umbelliferae)

Of all the weeds that plague the British gardener
GROUND ELDER or GOUTWEED, *Aegopodium podagraria*, is
one of the most difficult to eradicate. It *can* be got rid of
— by deep-digging and grovelling after each broken-
off bit of rhizome or underground stem (which at least
are white and easily seen), but only if constant vigil is
kept. A native of Eurasia, it was introduced into Britain
by the monks as a cure for gout. Bishops particularly
seemed to suffer from the complaint, so Bishop's Weed
it became. It was to St Gerard that sufferers appealed for
relief from 'joint-ache', and thus the plant earned the
name Herb Gerard. It is also called Jack-jump-about
because of the way the seeds scatter. From Britain it
spread to eastern Canada and the eastern United States.

Varying in height from 16 to 40in (40–100cm),
Ground Elder, with its hollow grooved stem, is
recognizable by its shining bright-green leaves, the
upper leaves being ternate (divided into 3) and each
stalked, the leaves springing from the roots being twice
ternate, all being irregularly saw-edged. The stem-
leaves grow opposite each other in pairs, and from their
axils soar the umbrellas of the inflorescences bearing
15–20 clusters of tiny white starry flowers, 5-petalled,
each of which will develop into a fruit containing only
2 seeds. The plant flowers from May to July. While it is
comforting to know that Ground Elder is an indicator
of a rich soil, take warning from the fact that its
creeping rhizomes will dominate 3 square yards in a
season.

GROUND ELDER or GOUTWEED

GROUNDSELS (Compositae)

GROUNDSEL or COMMON GROUNDSEL, *Senecio vulgaris*, was prized in the 15th-century English physic garden. The leaves, boiled in wine or water, made an infusion that healed the 'paine and ach of the stomacke that proceeds of Choler', as physician-herbalist John Gerard recommended. It had other healing virtues and in 1620 the Pilgrim Fathers took it to New England where by 1672 it had outrun its usefulness, for John Josselyn in his *New England's Rarities Discovered* includes it in his list of English immigrant weeds.

Groundsel is now widespread throughout the temperate regions and is one of the commonest weeds in gardens. No wonder! The average number of offspring from a single plant can be about 1,000, and when you consider that it flowers all the year round and that its life-cycle is a brief five weeks, this means a potential population of some 1,000–million new plants by the end of the third generation in the autumn — all from one plant. The seeds, each carried on a parachute of hairs, are effectively wind-borne. In wet weather they

GROUNDSEL or COMMON GROUNDSEL

become sticky and readily adhere to anything that touches them, so that boots and the feet of animals and birds are alternative means of transporting them to fresh growing-grounds. Wild birds enjoy their seeds and account for some destruction, though this is also a means of dispersal, for seedlings have been raised from their droppings.

Groundsel has a rather ragged appearance because of the alternate, pinnately cut and toothed leaves and irregularly branched stems which are weak and rather succulent. Its height varies between $3\frac{1}{2}$ and 18in (9–45cm), though it does not often grow so tall. The flowers are mere tufts of yellow florets above a tubular calyx-like structure formed of green bracts tipped with black. The leaves are softly hairy, and the root fibrous, which makes it easy to weed. Its destruction should be pursued ruthlessly, especially at the end of the year and if you live in farming country, for it is a host plant providing house-room through the winter for the aphis *Myzus persicae*. No farmer growing beet nearby will thank you for harbouring it: this greenfly is the vector of Beet Mild Yellowing Virus (BMYV), in the United States the similar Beet Western Yellows Virus. Gardeners, however, can take heart from the presence of Groundsel, as it is an indicator of nutrients and nitrogen in the soil.

STINKING GROUNDSEL, *Senecio viscosus*, is equally well-named STICKY GROUNDSEL because of its very viscid hairy stems varying in height between 4 and 24in (10–60cm). It is therefore a taller plant than the commoner Groundsel and, although the leaves are the same shape, the margins of their segments are not toothed and they are a dark green instead of a light green. Also the flower-heads are more open, with spread ray-florets, yellow and about 13 in number surrounding the flat yellow cushion of disk-florets. The flower-heads are long-stalked, the buds almost round. An annual flowering from July to September, locally common and scattered throughout lowland Great Britain, the Sticky Groundsel is increasing. It is widespread in Europe and is an introduced weed in North America.

STINKING or STICKY GROUNDSEL

HEMLOCK (Umbelliferae)

HEMLOCK, *Conium maculatum*, is a member of the Parsley family, but all parts of the plant are highly narcotic and poisonous. In the United States and Canada it is plainly labelled POISON HEMLOCK (Socrates died of it in the spring of 399 BC). It is an erect branched biennial up to 6 or 7ft (2m) tall with a stinking smell. The stems are furrowed, smooth, with a bluish tinge and purple-spotted. The leaves grow from the base and opposite each other on the stem. They have a lacy appearance, being soft in texture and bipinnate with toothed margins, and are wide and triangular in outline and up to 1ft (30cm) long. The sharp-tipped ends of the leaflets are light green or yellowish. It is the conspicuous waved ridges, pale brown, of the 2-seeded fruit which distinguish the Hemlock from any other similar plant. The thick tap-root is long, white, and often branched. Carried on branching umbels, the terminal inflorescence blooms first, in a mass of tiny white 5-petalled flowers, but is soon overtopped by the others. The flowering period is from June to September. A native of Eurasia, the Hemlock is naturalized throughout most of the British Isles and all the United States excepting an area between central Montana and north-eastern Minnesota, and throughout Canada.

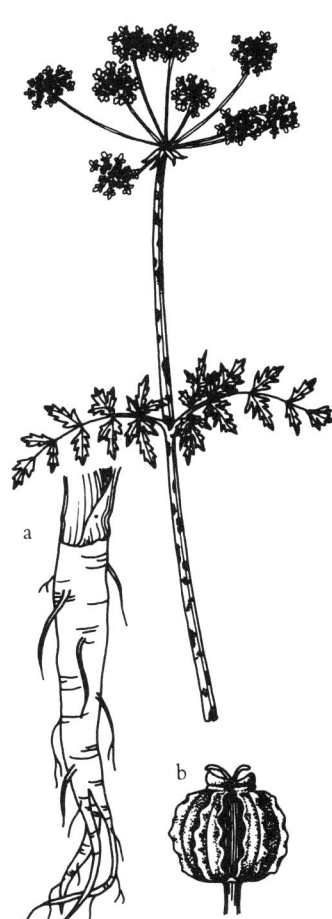

HEMLOCK or POISON HEMLOCK, flowering umbel *a*. Root *b*. Fruit

HEMP DOGBANE
(Apocynaceae)

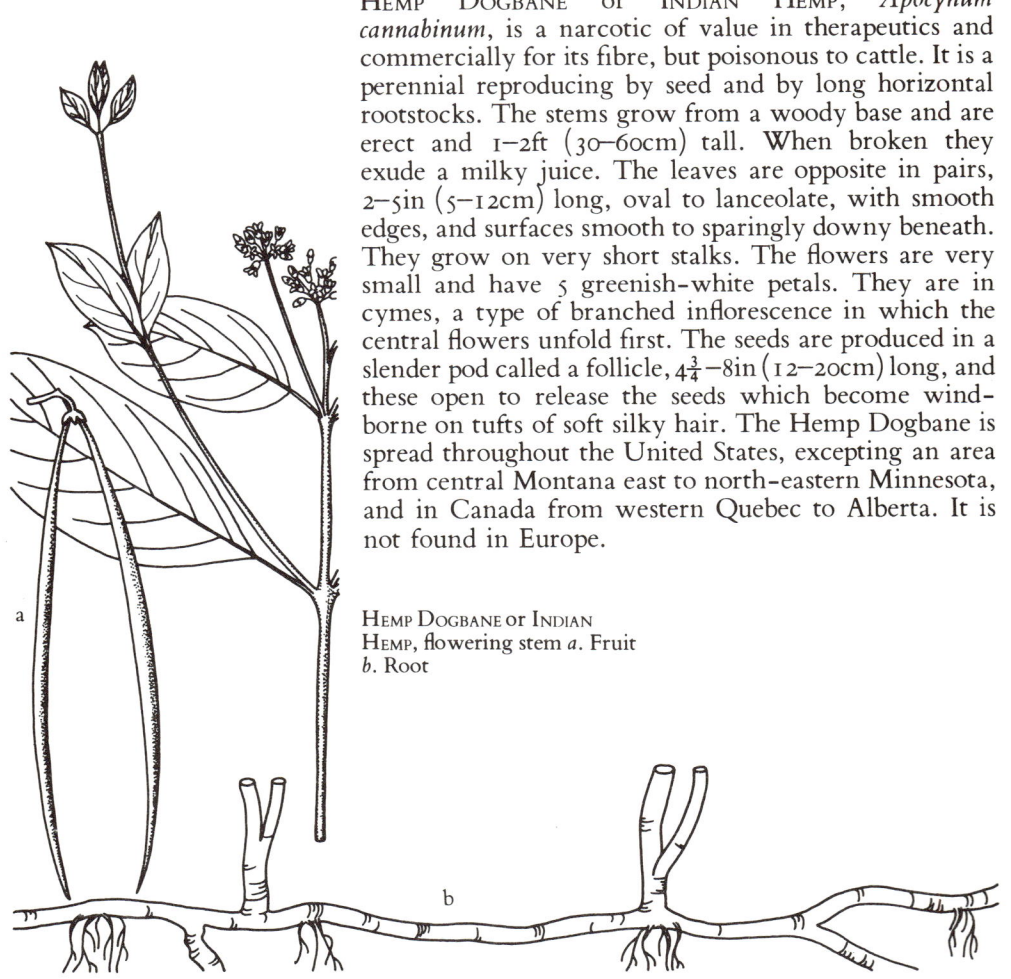

HEMP DOGBANE or INDIAN HEMP, *Apocynum cannabinum*, is a narcotic of value in therapeutics and commercially for its fibre, but poisonous to cattle. It is a perennial reproducing by seed and by long horizontal rootstocks. The stems grow from a woody base and are erect and 1–2ft (30–60cm) tall. When broken they exude a milky juice. The leaves are opposite in pairs, 2–5in (5–12cm) long, oval to lanceolate, with smooth edges, and surfaces smooth to sparingly downy beneath. They grow on very short stalks. The flowers are very small and have 5 greenish-white petals. They are in cymes, a type of branched inflorescence in which the central flowers unfold first. The seeds are produced in a slender pod called a follicle, $4\frac{3}{4}$–8in (12–20cm) long, and these open to release the seeds which become wind-borne on tufts of soft silky hair. The Hemp Dogbane is spread throughout the United States, excepting an area from central Montana east to north-eastern Minnesota, and in Canada from western Quebec to Alberta. It is not found in Europe.

HEMP DOGBANE or INDIAN
HEMP, flowering stem *a*. Fruit
b. Root

HORSETAIL (Equisetaceae)

Both Britain and America claim the COMMON or FIELD HORSETAIL, *Equisetum arvense*, as a native. It has other names: Mare's Tail, Pipeweed and Paddy's Pipe, the last two referring to the cup-shaped sheaths growing around the stem. There are two kinds of stem. In spring brownish stems develop, a sheath encasing the growing-point at each period of growth, so that the stem is punctuated upward at regular intervals with these cups. The final growing-point ends in a cone containing spores; and when the spores are shed, this stem withers away and the bushy green vegetative stems appear. These have whorls of bristles growing from beneath the sheaths. The plant has no leaves, for it is a relic of primeval times, the genus *Equisetum* being a link between modern plants and some of the most ancient orders of vegetation: our coal measures are mainly composed of the fossilized remains of immense forests of gigantic, woody, *Equisetum*-like plants which once covered the earth.

Another name for the Horsetail is Scouring Rush. The plant contains silica and in the 17th century was used to give a fine finish in scouring pewter. It is still used occasionally to produce a satin finish on cabinet work.

The spring-growing fertile stems grow to a height of 4–10in (10–25cm) and are unbranched; the sterile stems 8–32in (20–80cm). Because of its underground rhizomes, which are attached to small tubers, the Horsetail is a rampant spreader which many gardeners have found ineradicable. I have, however, heard of a cure: it is to sow the area with nasturtiums. These plants act as a smotherer and finally exterminator, for the Horsetail cannot tolerate overshadowing competition. It is poisonous to livestock, especially horses and cattle, causing equisetosis.

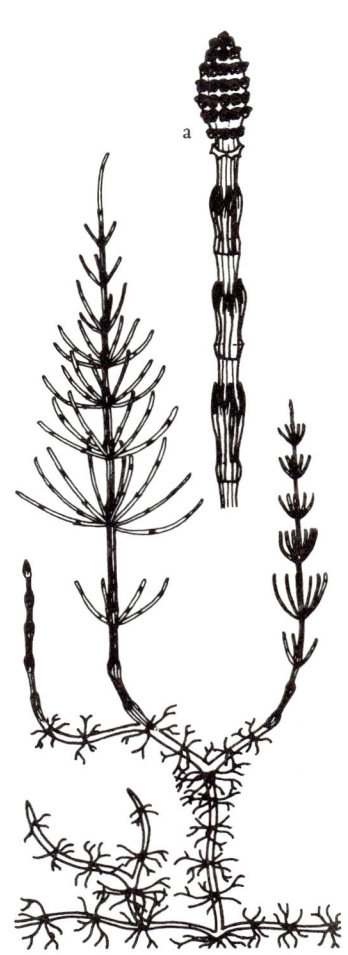

COMMON or FIELD HORSETAIL, with sterile stems *a*. Fertile stem

IVIES (Araliaceae, Labiatae, Anacardiaceae)

The three ivies to be described are grouped here not because they belong to the same family or genus but merely because they all bear this common name.

A European spread throughout the British Isles, the IVY, *Hedera helix*, the ENGLISH IVY of America where it is found in the open woods of Virginia and southward, has two kinds of leaves: those growing on the sterile branches having 3–5 more or less triangular lobes; the others, which grow on the flowering branches, having no lobes and being rhomboid. Ivy is a perennial densely clothing whatever it climbs on, or forming thick carpets if creeping along the ground. The leaves, which are alternate and evergreen, are smooth, dark green above, often with pale veins and sometimes tinged with purple, pale green beneath. In October it bears clusters of tiny 5–petalled green flowers ripening to green berries the following spring. Its method of climbing is by aerial roots which find their way into cracks and crevices. As a carpeter it uses these surface roots to anchor itself to the ground. It can be a menace in eating into the mortar of old walls, the weight of its foliage and branches eventually causing their destruction. On the ground its dense growth of impermeable leaves will prevent rain from seeping down into the soil.

One good word for the Ivy: it can be useful in densely shady places where, without being asked, it will take over a straggling old hedge that has ceased to thrive, providing a substantial windbreak and nesting places for various birds, your allies in helping to keep down insect pests. But always beware of seedlings growing where they are not wanted. Ivy is a determined grower.

GROUND IVY, *Glechoma hederacea*, also known in North America as GILL-OVER-THE-GROUND and CREEPING CHARLIE, is a delightful weed that — provided it can be kept in check, which is not difficult — is no harmful intruder unless it gets into lawns. It belongs to the Labiatae family, and its lipped flower of bright violet and round crinkled leaves are both attractive. Indeed, it

IVY or ENGLISH IVY, fertile stem with fruit *a*. Creeping stem with aerial roots

was once a treasured little garden plant. A perennial, it has two types of stems, one that creeps along the ground and roots at the nodes, the other sending up erect flowering stalks. The leaves are also stalked, opposite, with crenate margins. The flowers, $\frac{1}{2}$–1in (13–25mm) across, are usually in pairs in the axils of the upper leaves. A native of Eurasia it is common throughout the British Isles except in northern Scotland, and throughout North America. The flowering period is from April to June.

For obvious reasons, a weed not to be encouraged is the POISON IVY, *Rhus radicans*, which belongs to the Anacardiaceae family. The juice of every part contains a poison that causes an irritating inflammation and blistering of the skin. A perennial woody shrub or vine, the leaves are alternate and compound with three large, shiny, stalked leaflets very variable in shape and in the toothing of the margins. As a shrub, which is its most frequent type, it grows 12–18in (30–45cm) high. As a climber it produces aerial roots to ascend posts or trees. The small yellowish-green 5-petalled flowers in June and July are in dense clusters on long stalks from the axils of the leaves. These are followed by small whitish or yellowish fruit somewhat resembling a peeled orange. As well as reproducing by its seeds, the Poison Ivy does so by creeping rootstocks from the basal stem nodes, sometimes running horizontally underground for several metres and sending up leafy shoots from their nodes.

This poisonous weed is native to America and is distributed throughout the United States, north into southern Canada from Quebec to British Columbia, south into Mexico and the West Indies. Britain and continental Europe are spared its presence.

POISON IVY *a*. Stem with aerial roots

JIMSON WEED (Solanaceae)

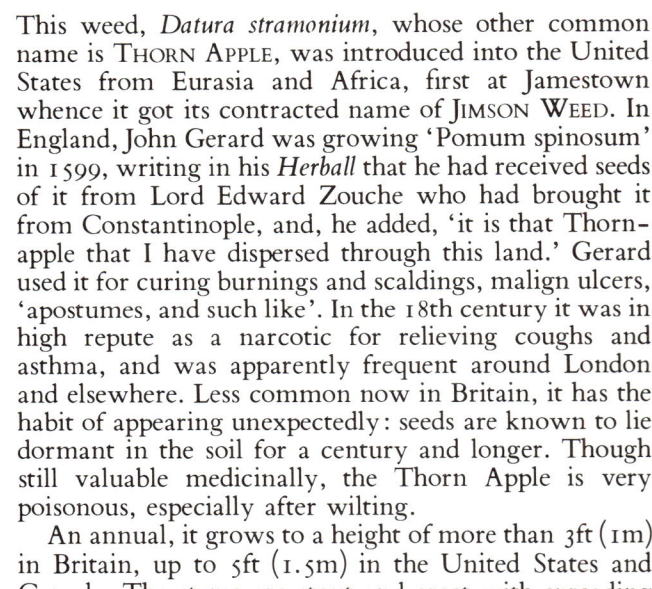

This weed, *Datura stramonium*, whose other common name is THORN APPLE, was introduced into the United States from Eurasia and Africa, first at Jamestown whence it got its contracted name of JIMSON WEED. In England, John Gerard was growing 'Pomum spinosum' in 1599, writing in his *Herball* that he had received seeds of it from Lord Edward Zouche who had brought it from Constantinople, and, he added, 'it is that Thorn-apple that I have dispersed through this land.' Gerard used it for curing burnings and scaldings, malign ulcers, 'apostumes, and such like'. In the 18th century it was in high repute as a narcotic for relieving coughs and asthma, and was apparently frequent around London and elsewhere. Less common now in Britain, it has the habit of appearing unexpectedly: seeds are known to lie dormant in the soil for a century and longer. Though still valuable medicinally, the Thorn Apple is very poisonous, especially after wilting.

An annual, it grows to a height of more than 3ft (1m) in Britain, up to 5ft (1.5m) in the United States and Canada. The stems are stout and erect with spreading branches, smooth, green or purple. The leaves, which are alternate, are dark green and strongly scented, oval to triangular with widely toothed margins and ending in a sharp point. Due to their habit of folding together they have a prickly appearance.

Distinctive are the large white 5-lobed trumpet flowers, $2\frac{1}{2}$–3in (4–7.5cm) across, growing in the axils of the branches and followed by oval green fruits beset with short sharp spines, though sometimes unarmed. Each fruit may contain between 400 and 800 seeds, and germination is normally high.

The Jimson Weed is spread throughout all the United States excepting a large area in the north-western and north-central parts and an area in south Texas. It inhabits a distinct area in Colorado and is found scattered across Canada. It flowers from June to October.

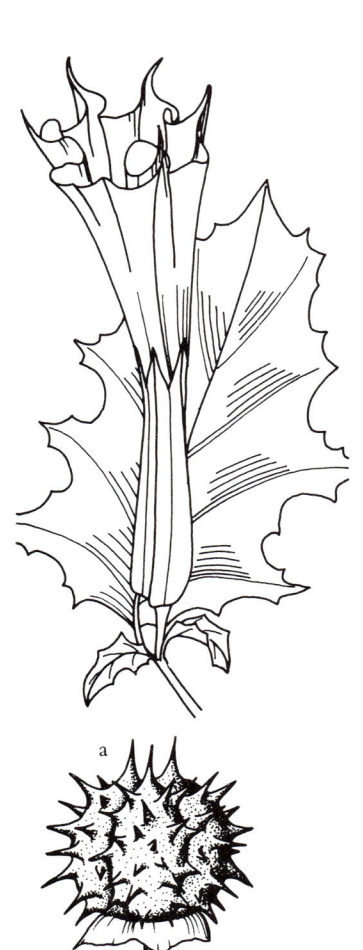

JIMSON WEED or THORN APPLE
a. Fruit

LETTUCES (Compositae, Valerianaceae)

The BLUE LETTUCE, *Lactuca pulchella*, is a perennial and a native of western North America, given its common name from the colour of its ray-florets which may, however, also be purple. The flower-heads, which are in clusters, are nearly 1in (2.5cm) across, and the flowering time is usually July to September, in the Pacific states June to October. It grows 1–3ft (30–90cm) tall. The lower leaves are thickish with downward-pointing projections like two pairs of triangular teeth, the main part of the leaf extending to a long triangle and narrowing at the stem to a winged stalk; the upper leaves are mostly unlobed and narrowly lanceolate. All the leaves are alternate. The achenes, single-seeded fruits, which are dark red-brown and sometimes slate-grey, are dispersed by means of parachute tufts of white hairs. The plant also increases from a deep tap-root that sends out long horizontal roots producing new shoots.

The Blue Lettuce is common throughout an area extending from central Washington east to the Great Lakes, and south as far as the Mexican border in New Mexico, with a distinct area in Oklahoma. It is rare in the north-eastern states and rare to common in the eastern provinces of Canada.

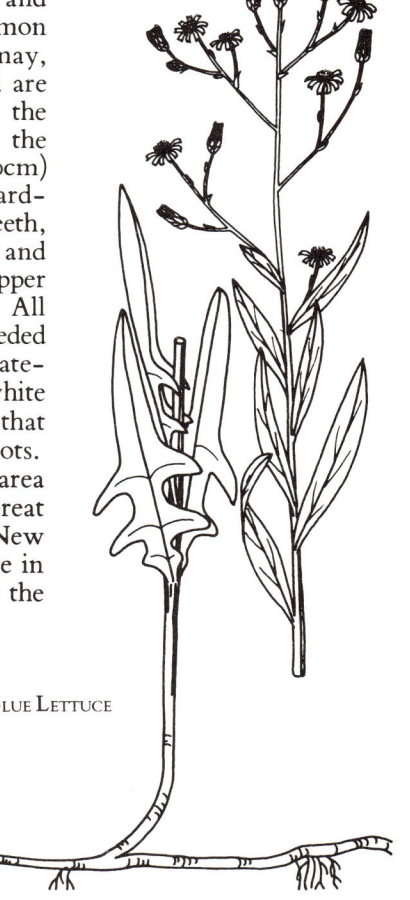

BLUE LETTUCE

The European *Lactuca virosa* is not one for the salad bowl, for it is the POISONOUS LETTUCE or ACRID LETTUCE. An annual or more usually a biennial, it is a sub-Mediterranean species flowering in July and August. It can attain a height of over 6ft (2m). Although recorded as being found in England in the first half of the 17th century it was uncommon until the third decade of the present century when, with the extensive construction of arterial roads offering suitable habitats

Leaf of POISONOUS LETTUCE

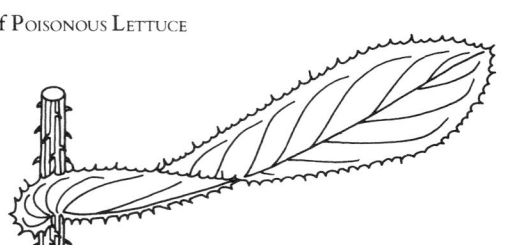

on unclothed embankments, the Poisonous Lettuce soon spread. The leaves are oval with prickly saw-edged margins, the lower part wrapping around the stems which are whitish or reddish and sometimes prickly under the leaf. The lower leaves may be undivided or deeply cut. All the leaves are alternate and prickly on the underside of the main veins. The inflorescence is branched, the pale greenish-yellow composite flowers borne on branchlets growing almost at right angles to the main flowering stem, each flower-head being $\frac{2}{5}$ in (10mm) across.

The PRICKLY LETTUCE, *Lactuca serriola* (*scariola*), is an over-wintering or biennial weed. Introduced into North America from Europe it has become a pest and is dangerous to cattle. In Britain it confines itself to England and Wales. It grows from 1ft tall up to 4 or 5ft (30–150cm) and is much like the Poisonous Lettuce in appearance. But there is one obvious difference: the stem-leaves when fully exposed to the sun all stand with their blades vertical in the north-south plane, giving it the nickname of 'Compass Plant'. The leaves are alternate on the stem, oval in outline and either undivided or deeply divided, the upper leaves unlobed, but all with back-curling lobes at the base. A native of Eurasia, the Prickly Lettuce has spread throughout all the United States excepting areas in extreme northern Maine and extreme southern Florida. In Europe it

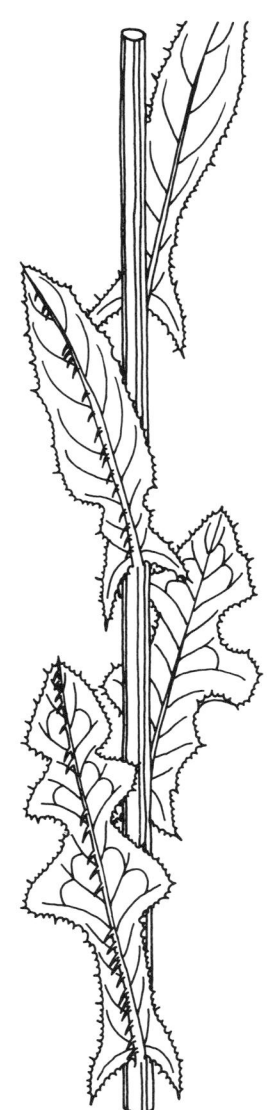

PRICKLY LETTUCE

extends across the southern and central regions, the Netherlands, north-west Germany, Denmark and Gotland. It flowers from July to September.

LAMB'S LETTUCE or CORN SALAD belongs to a different family, Valerianaceae, its botanical name being *Valerianella locusta (olitoria)*. This is the little over-wintering annual weed that looks like a tiny Forget-me-not with lilac-blue flowers. These are funnel-shaped with 5 round lobes. The plant is edible and was

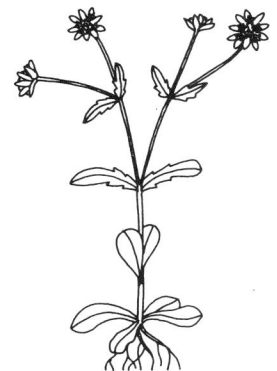

LAMB'S LETTUCE

much used as an early spring salad in past centuries. Growing erect to a height of 3–16in (7–40cm), the stems are rather brittle, much-branched and slightly downy below. The leaves grow in pairs opposite each other, the lower ones being lanceolate, the upper ones oval and narrower at the top of the stem, with winged lobes at the base. The little clusters of flowers are surrounded by tiny leafy bracts, so that they resemble a Victorian posy. Flowering in April and through June, Lamb's Lettuce is common throughout nearly the whole of the British Isles and southern Europe; in the United States from New England to Indiana, south to North Carolina and Tennessee, and the western states.

LIPPED PLANTS (Labiatae)

The flowers of the Labiates are irregular: that is, with the petals not arranged symmetrically as they are in the Cruciferae or Cress family where, if you took a knife and cut a flower squarely into four, each quarter would be the same. Lipped plants are tubular and 5-lobed, most of them having a hooded top lobe arching over and serving to protect the pollen on the stamens. Below the open mouth of the flower, as it were, is a lobed lip with well-marked honey guides. The stems are always quadrangular and the leaves always opposite. It seems as if the flowers, which grow in the axils of the upper leaves, are arranged in whorls around the stem: they are rarely solitary. In fact they grow in two opposite cymes (branched panicles); but so closely are they clustered that they appear to be a whorl of up to 10 or more flowers. Most of the species have a peculiar strong scent, either highly aromatic (thyme, mint, sage, and many other pot-herbs belong to this family) or disagreeable as in plants like the Hedge Woundwort (*Stachys sylvatica*).

The hairy perennial WHITE DEAD-NETTLE, or SNOWFLAKE, *Lamium album*, which is not a stinging nettle, is unbranched, 8–24in (20–60cm) in height with creeping rhizomes and erect stems. The heart-shaped leaves are stalked and coarsely toothed. Though opposite, each pair lies at right angles to the pairs above and below, marking as it were the four main points of the compass. The clear white flowers are quite large, 1in (2.5cm) long, the lateral lobes of the lower lip having 2 or 3 small teeth. They bloom from March to November. Common in England, the White Dead-nettle is rare in the north of Scotland. It ranges from Scandinavia to central Spain, Italy, Macedonia and farther east, in North America locally in old lawns. Other names for it are the Bee Nettle and Honey Flower, telling of its attraction to long-tongued bumble-bees; and if you hold a flower upside-down so that the two lower stamens, black-headed and golden, are lying side by side, you will see why it is also called Adam and Eve. Its other name, Archangels, refers to the clusters of flowers supposedly resembling a choir of

WHITE DEAD-NETTLE or SNOWFLAKE

white-robed figures, and this name is applied also to *Lamium purpureum* and to the European *Galeobdolon luteum*, Yellow Archangel.

The RED or PURPLE DEAD-NETTLE, *Lamium purpureum*, is very common throughout the British Isles, but in continental Europe found only in the mountains in the south. It was introduced into Pennsylvania. Growing 4–18in (10–45cm) high, it branches from the base and is softly hairy with pinkish-red flowers often tinged with purple, and the stems often red-purple. The leaves, which are stalked, are roundish or oval, heart-shaped at the base, the margins being evenly crenate (round-toothed). The Red Dead-nettle is a persistent invader of gardens, its presence indicating nutrient-rich and mildly humic, loamy or sandy-loam soils. The flowering period is from April to late autumn and often throughout the winter. Each plant produces about 200 seeds.

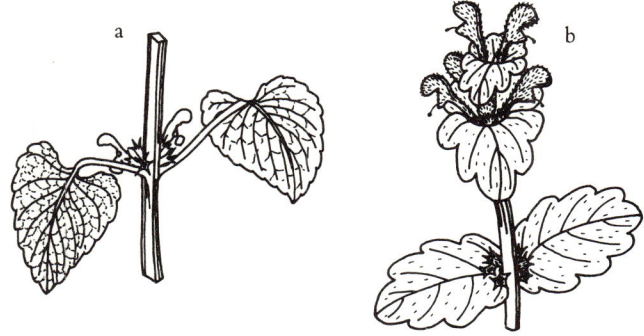

a

b

Flowers and leaves of *a*. RED or PURPLE DEAD-NETTLE
b. HENBIT

HENBIT, *Lamium amplexicaule*, is common all over Britain, North America and Europe. An annual with branches ascending from the base, its height is 2–10in (5–25cm). The leaves are a quite different shape from those of the first three species, being almost round and deeply crenate, the lower leaves long-stalked, the upper ones unstalked, each pair together making a ruff around the stem, hence the epithet *amplexicaule*, stem-clasping. The tube is clothed with white down and it and the lower lip of the flower are pale pink surmounted by a crimson hood. The flowering period is from March to August but it can bloom again later and even the whole year round. All the Archangels contain useful elements for the compost heap.

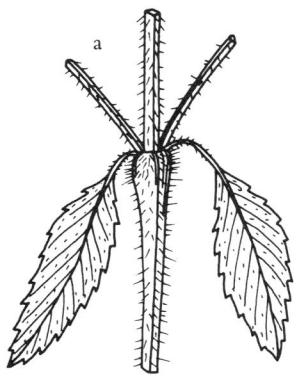

Stems and leaves of
a. COMMON HEMP NETTLE
b. BUGLE

Another pink-flowered Labiate that might at first glance be confused with other members of the family is the annual COMMON HEMP NETTLE, *Galeopsis tetrahit*. An easily recognizable distinction is that the stems are swollen below the leaf-junctions. Very prominent also is the tuft of red-tipped glandular hairs where the flowers arise. The height of the plant varies considerably from 4 to 40in (10–100cm). The leaves are oval and pointed and somewhat hairy. Of an overall pale pink, the lower lip of the flower, which is divided into 3 lobes, has clearly marked honey guides which do not reach the margin. A native of Eurasia, the Common Hemp Nettle is found throughout roughly the north-eastern two-thirds of the United States, and in Canada from Newfoundland to British Columbia and Alaska. It is common all over the British Isles and ranges from Iceland and the Faroes to European Russia, and south to central Spain, Montenegro and Macedonia. The flowering period is from June to September.

The BUGLE, *Ajuga reptans*, is a perennial 4–12in (10–30cm) tall with lilac-blue flowers lacking the usual hooded top lip, so that the stamens and stigma are exposed. The specific epithet *reptans* means creeping, and this is exactly what the Bugle does, using its long leafy stolons (rooting runners) to spread itself. The basal leaves form a rosette, the lower stem-leaves being drawn out at the base into a long stalk, the upper pairs almost stalkless. All the leaves are smooth, oval, and with crenate margins. The square stems are unbranched and often hairy on two opposite sides. The Bugle is a European ranging from Scandinavia and northern Russia to central Portugal, Sicily, Greece and the Caucasus, south-west Asia, eastern Algeria and Tunisia. It is common throughout the British Isles but has not travelled to North America.

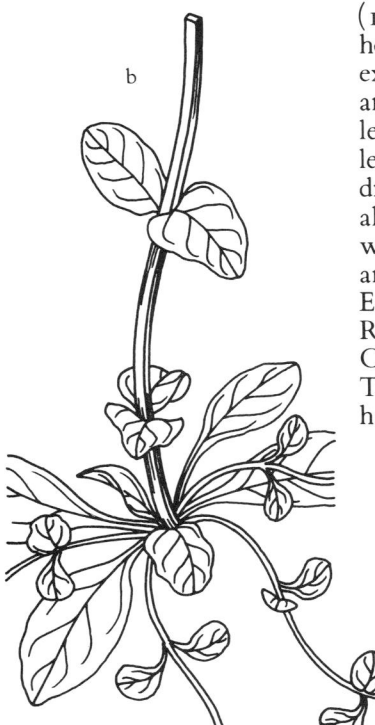

SELF-HEAL, *Prunella vulgaris*, America's HEAL-ALL, has a stubby head of bright purple flowers in whorls of six, always with a pair of leaves under it. It is a perennial with short rhizomes and erect stems 2–12in (5–30cm) high. The stalked leaves are oval and pointed, the margins not toothed. Flowering from May to September, Self-heal is very common throughout the British Isles, southern Europe and all the United States, excepting an area between central Montana and eastern North Dakota; in Canada from Newfoundland to British Columbia and Alaska. Akin to it is the LARGE SELF-HEAL, *Prunella grandiflora*, a native of continental Europe. In this species, larger in every respect, the flower-heads have no leaves at the base.

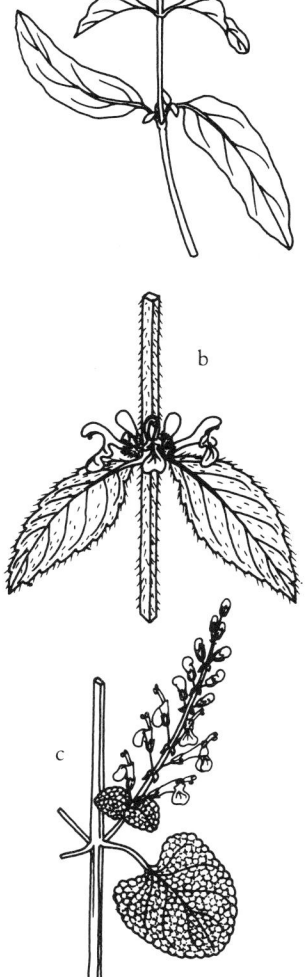

HEDGE WOUNDWORT, *Stachys sylvatica*, is a perennial with long creeping rhizomes which are green, coarsely hairy and with a disagreeable smell when bruised. The stems are hollow, simple or slightly branched, growing to a height of 12–40in (30–100cm). The bright claret-red flowers are in whorls of 6 forming an open spike, the lower lip of the flower being 4-lobed and having a honey-guide between the two lateral lobes in the form of a cross on a white background. The leaves are heart-shaped, toothed and stalked. In flower from June to October, the Hedge Woundwort, also called the Wood Woundwort, is common throughout the British Isles and ranges from southern Norway to north-east Portugal, central Spain, Sicily, Albania and Thrace. It is found locally in New York and Pennsylvania.

Easily recognizable among the lipped plants is the WOOD SAGE, *Teucrium scorodonia*, by its branched clusters of yellowish-green flowers growing in pairs, arising from a single axis with a pair of leaves below. The stems are reddish and the leaves much wrinkled, heart-shaped and stalked on the stem. A perennial with creeping rhizomes, the Wood Sage has erect stems growing to a height of 6–12in (15–30cm). It is common in Great Britain and is distributed throughout Europe except in the extreme north; it is locally established in Ontario and Ohio. It flowers from July to September.

Flower-head of *a*. SELF-HEAL or HEAL-ALL *b*. HEDGE WOUNDWORT *c*. WOOD SAGE

LORDS-AND-LADIES
(Araceae)

LORDS-AND-LADIES

The European *Arum maculatum*, LORDS-AND-LADIES or CUCKOO-PINT, is one of the plants that fascinated Charles Darwin because of its partnership with insects and the complexity of its floral mechanisms for procuring pollination. In the flowering months, April and May, it is easily recognized by the hooded spathe that enfolds the purple poker-like spadix, a structure heating up like an electric rod to emit the smell of rotting meat which attracts its pollinators, tiny dipteran flies and owl-midges. The insects cannot resist this smell, but alighting on the spathe — oily and slippery as a greasy floor — they are sent chuting down through a ring of stiff hairs which prevents their escape. Below is a ring of male flowers, and around the bottom a ring of female flowers on which the midges crawl about, finding drops of nectar. While they are feasting, the male flowers above them rain down a shower of pollen. The trapped insects are not released until the following morning, by which time the bristly barricade has softened, allowing the pollen-dusted midges to crawl through to freedom — only to find another prison in the next *Arum*, but this time to leave the pollen on the ring of female flowers.

The Cuckoo-pint develops a cob of bright green berries that turn scarlet. These are poisonous and should be removed if children are about. The leaves, appearing in early spring, are long-stalked and arrow-shaped with a dark green midrib, often spotted with black. The plant grows from an acrid white tuber, and a fresh one is produced from the base of the stem each year. If tolerated, the Cuckoo-pint can become a pest. It is shade-loving, so look for it in out-of-the-way corners. Distribution is general throughout the British Isles and Europe as far north as the south of Sweden.

MALLOWS (Malvaceae)

The COMMON MALLOW of the British Isles and Europe, and the COMMON MALLOW of the United States, are two different plants, so they must be described separately. Both are weeds common to Europe and North America and are of European origin.

Malva sylvestris, the COMMON MALLOW of Britain and continental Europe, the HIGH MALLOW of America, is a large plant, 19–36in (49–90cm) tall but of a sprawling habit. The stems are woody below, branched and very hairy, the 5-lobed alternate leaves 2–4in (5–10cm) across, somewhat folded from the midrib, roughly triangular in outline, and often with a small dark spot. The 5-petalled flowers of a bright rose with prominent stripes of a deeper tint are 1–1½in (2.5–4cm) across and grow in clusters. This is a perennial flowering from June to September.

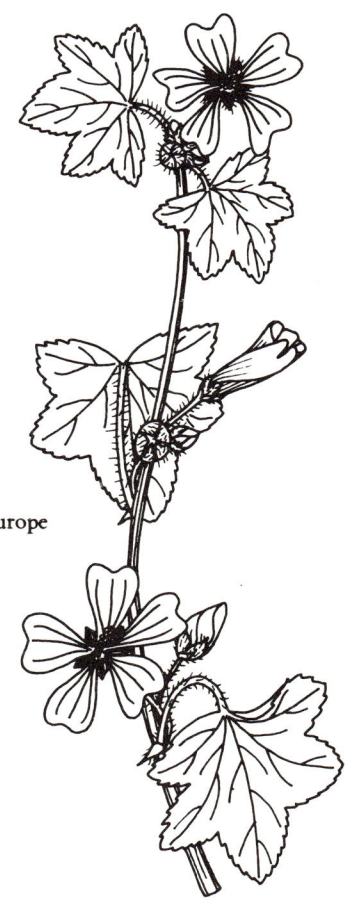

COMMON MALLOW of Europe
or HIGH MALLOW

Leaf and flower with fruit of DWARF MALLOW or COMMON MALLOW of United States

The COMMON MALLOW of the United States is *Malva neglecta* (in Britain called the DWARF MALLOW), a

122

prostrate annual or biennial, in height 6–24in
(15–60cm) and with a short straight tap-root. The 5
notched petals are a pale pink, veined, and only $\frac{3}{4}$–1in
(2–2.5cm) across. The leaves, alternate and 5-lobed but
rounded, are $1\frac{1}{2}$–$2\frac{3}{4}$in (4–7cm) across and on long
stalks. The flowering period in Britain is June to
September, in the United States April to October.
Canada's name for the plant is ROUND-LEAVED
MALLOW. Its botanical synonyms, that is the Latin
names by which this Mallow was known previously,
are *Malva rotundifolia* and *Malva vulgaris*, but it is one
and the same plant as is mentioned by John Josselyn as
having reached New England before 1669. Today it is
one of America's commonest and most pernicious
weeds, though children eat its flat seeds, calling them
'cheeses', and the leaves are often used for garnishing
dishes.

Abutilon theophrasti, the VELVET-LEAF, was introduced
into the United States from India as a garden plant. It
escaped to colonize all the states excepting mainly a
large area along the northern boundary, and spread into
all the eastern provinces of Canada. Its common name
describes the soft surface of its leaves which are large
and heart-shaped, sharply pointed and growing
alternately up the branching stems. The 5-petalled
yellow flowers, up to 1in (2.5cm) across, are on short
stalks in the leaf-axils. An annual, 2–4ft (60–120cm)
tall, it has a thick tap-root. Other names for it are
Butterprint and Pie-marker, referring to the pattern
made by the fruit, a circular cluster of 12–15 beaked
pods, each about 1in long and containing several seeds.
The flowering period is August and September.

Leaf and flower with fruit of
VELVET LEAF

MAYWEEDS and CHAMOMILES (Compositae)

These are usually strongly-scented weeds with daisy flowers sometimes yellow, sometimes white or pink, and leaves so much divided and sub-divided that they are feathery in appearance. They grow alternately on the stems.

The CORN CHAMOMILE, *Anthemis arvensis*, is a bushy hairy annual 5–20in (12–50cm) in height. The flower-heads, which are long-stalked, solitary and 1–1½in (2.5–4cm) across, grow from the axils of the leaves. The white ray-florets are female (with styles). Each plant yields about 4,000–5,000 seeds, and as these germinate on the surface of the soil the moral is — don't let it reach that stage. It likes a soil that is mineral-rich, lime-free and usually strongly acid loamy and sandy loamy.

Locally common throughout the British Isles, the Corn Chamomile is spread over Europe northward to the south of Norway, central Sweden and across to Asia Minor. In North America it was first introduced into New England and New York and now inhabits the United States west to Illinois and the Pacific area, as well as eastern Canada. It flowers in June and July.

Similar in habit is the STINKING MAYWEED, *Anthemis cotula*, which lives up to its English common name and its American one of STINKING CHAMOMILE, otherwise MAYWEED. It differs in having stems that are taller, 8–24in (20–60cm), and solitary but often twin flowers that are slightly smaller, ½–1in (1.2–2.5cm) across. Also, the ray-florets are usually neuter (without styles), white as in the Corn Chamomile but the petals later turning under. The much-divided leaves are sparsely hairy. In Britain it is a locally common weed especially in south and central England, distributed in Europe from the south northward to southern Norway, and throughout the whole of the United States and Canada. It is an indicator of a loam soil. In Europe the flowering period is July to September, in North America May to October.

CORN CHAMOMILE

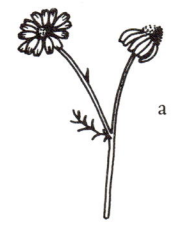

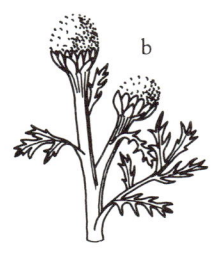

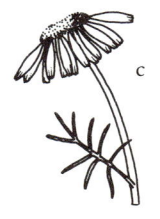

Flower and leaf of *a*. STINKING
MAYWEED or MAYWEED
b. RAYLESS MAYWEED or
PINEAPPLE WEED *c*. SCENTLESS
MAYWEED or
SCENTLESS CHAMOMILE

Know the RAYLESS MAYWEED, *Matricaria matricarioides*, by the absence of ray-florets, the solitary stalked flower-heads growing alternately up the stem consisting merely of disk-florets, a small round button of greenish-yellow held in a cup of sepal-like bracts, green edged with white. An aromatic annual of sturdy growth it is more commonly known as the PINEAPPLE WEED because of its scent. Its height varies between 2 and 16in (5–40cm), the stems being smooth and with many rigid branches. Flowering from May to November, the Pineapple Weed is common throughout Europe and North America. It is a rapid spreader, each plant producing about 5,300 seeds which again germinate on the surface of the soil. It likes open nutrient-rich loamy and sandy-loam soils.

If the Pineapple Weed is troublesome, the SCENTLESS MAYWEED or SCENTLESS CHAMOMILE, *Tripleurospermum maritimum* (*Matricaria maritima*), is ten times more so, each plant producing a horrifying total of about 34,000 seeds, or, to put it more accurately, 10,000 to 210,000 in a well-grown specimen, these also germinating on the soil surface. For all its fecundity it has so far troubled American gardeners only near ports, otherwise confining itself to northern and central Europe and seaside districts of Britain. As its common name tells us, it is scentless, or almost so (a sub-species, *inodorum*, has no scent at all). An annual to perennial weed, it is variable in habit, growing erect, decumbent or prostrate, usually branching and with smooth stems 4–24in (10–60cm) long. The solitary flower-heads of white ray-florets and yellow disks flattened across the top — a distinguishing feature — vary in size from ½in to 2in (1.5–5cm), and grow on long stalks. The flowering period is from May to October.

Pernicious weed! whose scent
the fair annoys,
Unfriendly to society's chief joys.

William Cowper *Conversation*

MIND-YOUR-OWN-BUSINESS (Urticaceae)

The Latin name for this plant is *Helxine soleirolii*, its other common name Mother-of-thousands, and nothing could better describe the multitide of tiny leaves that swarm in a dense evergreen ever-spreading mat — over the ground, over stones, and more particularly into the crevices of stonework and brickwork. It is certainly the pestiest of pests. Knife it out, it will soon appear in the same spot. The only treatment is a selective weed-killer that will be absorbed by the leaders poking their way where they cannot otherwise be reached.

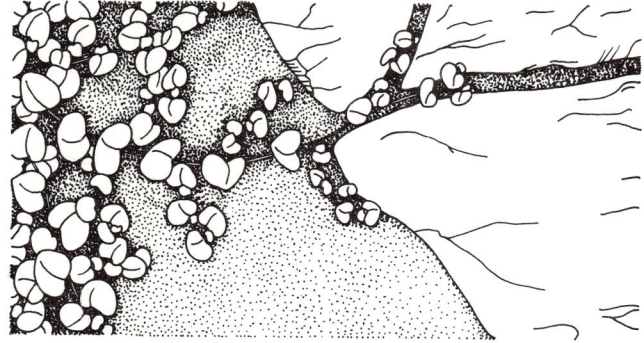

MIND-YOUR-OWN-BUSINESS

A perennial, MIND-YOUR-OWN-BUSINESS has slender stems 2–8in (5–20cm) long, the major measurement only if it finds really luscious damp soil in semi-shade. The leaves are $\frac{1}{8}$–$\frac{1}{2}$in (2–6mm) long, but are almost round, and grow alternately on the stem. You will not notice the tiny pink flower unless you take a close look from May to August: they are 4-lobed, growing singly from the leaf-axils, and surrounded by a leafy structure of one bract and two bracteoles called an involucre.

The one good thing about MYOB is that although it belongs to the Nettle family, it does not sting. It is an introduced plant, still used in Britain, America and continental Europe for carpeting rock gardens and cool greenhouses. Don't. Its homeland is Corsica. *Helxine* means 'Baby's Tears', and *soleirolii* comes from Joseph François Soleirol who made vast collections of Corsican plants in the 19th century. Would that he had left MYOB where it was!

MUGWORT (Compositae)

A cousin, as it were, of the Chrysanthemums and as strongly aromatic, the MUGWORT, *Artemisia vulgaris*, looks entirely different. Although belonging to the same Daisy family, its flower-heads never open in this way, the reddish-brown florets being mere tufts above a round calyx-like structure (involucre). And the flower-heads, instead of being at the top of the stem, are crowded on branches growing alternately up the stem. The stem-leaves, however, are the same shape, bipinnate and growing close to the stem alternately, the lower leaves short-stalked. They are dark green above and downy beneath. The grooved and angled stems vary in height between 2 and 4ft (60–120cm).

The Mugwort is common throughout the British Isles and most of the temperate regions of the northern hemisphere. In Canada it is commoner in the eastern part; in the United States throughout the north-eastern quarter, and along the Pacific coast from Washington through central California. It flowers from July to September.

MUGWORT, inflorescence and
stem-leaf

MUSTARDS (Cruciferae)

The seven weeds here grouped as Mustards all have this common name, although most of them belong to different genera. But they are all Crucifers, the flowers having 4 petals arranged like a cross. All are coarse-leaved and all are annuals with mustard-yellow flowers, excepting the Garlic Mustard which is a biennial with white flowers.

The CHARLOCK or WILD MUSTARD in Britain is *Sinapis arvensis*, in the United States *Brassica kaber* var. *pinnatifida*, by which last Latin name it is also known in Canada under the common name of COMMON MUSTARD. The plant is a European distributed throughout continental Europe and Britain, naturalized in the United States, in Canada common in agricultural areas. It has a slender tap-root and an erect single or branched stem, in Britain 12–30in (30–75cm) tall, in America growing to 9ft and over (nearly 3m). The stem is usually stiff and hairy, the leaves roughly hairy, the lower ones being stalked with a large very coarsely toothed terminal lobe, the upper leaves clasping the stem and alternate, usually simple, lance-shaped and coarsely toothed. The flowers grow alternately up the stem and develop pods, at the top of the stem growing in a cluster. Each plant produces about 1,200 seeds which can live in the soil for many years. The flowering period is May to autumn and often into the winter.

As the Mustards are all very alike, one must look for even the slightest distinction. The WILD RADISH or WHITE CHARLOCK, *Raphanus raphanistrum*, much resembles the Charlock, which has sepals spreading or drooping: those of the Wild Radish remain erect and are bristly. It has another feature distinguishing it from the Charlock: the pods on drying become much constricted between the seeds, so that the latter, 4–8, are plainly discernible; and the beak at the end of the pod is much longer. This weed's other name of White Charlock is because the clustered flowers, though usually yellow, can be that colour. (Or, to make it more difficult, they can be lilac, because the purple veins often suffuse the petals with this tint.) Other names for it are Runch, Jointed Charlock, and Kedlock. It blooms from May to September and grows 1–2ft (30–60cm) tall. Common throughout the British Isles and continental Europe, the Wild Radish is locally common in the eastern maritime provinces of Canada, becoming rare from Quebec to Alberta, more frequent again in British Columbia. In the United States it extends south to West Virginia and west to the Dakotas, Washington and Oregon.

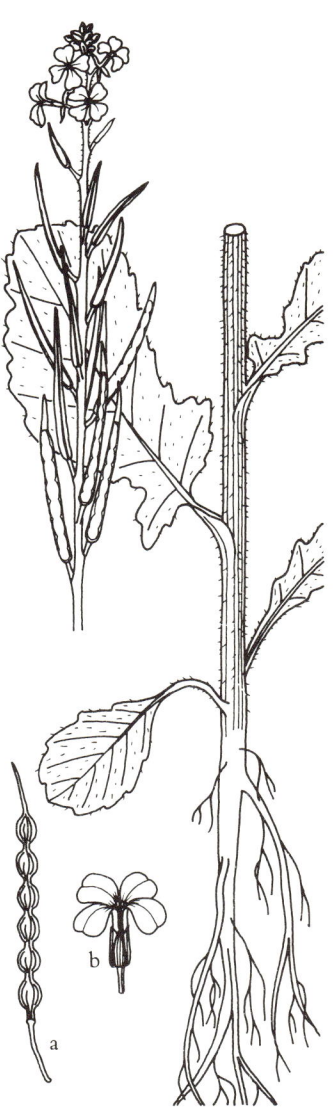

CHARLOCK or WILD MUSTARD, inflorescence and lower part of plant

a. Pod *b.* Flower of WILD RADISH

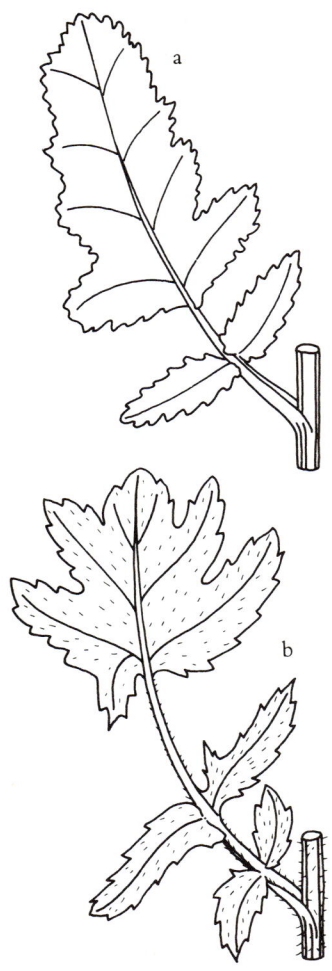

Leaf of *a*. BLACK MUSTARD
b. HEDGE MUSTARD

The BLACK MUSTARD, *Brassica nigra*, is a much taller weed with an erect shoot about 3ft (1m) or slightly more in Britain, 4–6ft (1.2–1.8m) in America. The leaves are very different, though all stalked, the lower ones having a large terminal lobe with two much smaller ones below, the alternate stem-leaves being lanceolate with only two tiny lobes, and no lobes on the topmost leaves. Again the flowers grow alternately up the stem and are clustered at the top, the pods being dense and erect. An Eurasian, the Black Mustard flowers from May to July and fruits from June to October. It is common in England and Wales, in Scotland only in the south, in Ireland in the south and east; widespread in central and southern Europe and throughout all the United States excepting an area between central Montana and eastern Minnesota. In Canada it is found locally.

This is the source of table mustard, which is made from its seeds. These also yield an oil used in medicine and soap-making.

The HEDGE MUSTARD, *Sisymbrium officinale*, has its basal leaves in a rosette, these deeply cut into paired lobes with a larger terminal lobe, all lobes being toothed. The stem-leaves are alternate with a long arrow-shaped terminal lobe and 1–3 small oblong lateral lobes. The clustered flowers are short-stalked with petals half as long again as the sepals, developing into stiffly erect pods with a rounded base. Both stems and leaves are bristly. It grows 1–2ft (30–60cm) tall in Britain, 2–3ft (60–90cm) in North America. May to October is the flowering period, and each plant yields about 2,700 seeds which germinate shallowly. A native of Europe, the Hedge Mustard is naturalized in North and South America, in Canada scattered throughout the eastern provinces and British Columbia, rare or absent on the prairies.

Sisymbrium altissimum is the TALL ROCKET in Britain; in North America it is a tumble-weed breaking off at the base when mature, to be bowled along the ground by the first wind and scattering its seeds as it goes. Its common name both in the United States and Canada is TUMBLE MUSTARD. The erect branched stem ranges in height from 8 to 40in (20–100cm) in Britain, in Canada up to 4ft (1.2m), though smaller plants are common, and in the United States up to 2ft (60cm). The basal leaves, which die before the flowering period (June to

August in England, May to September in America), are stalked, roughly hairy and with 6–8 pairs of narrowly triangular lobes; the middle stem-leaves, which are alternate, having more narrowly-shaped lobes; the uppermost leaves, also alternate, having lobes almost threadlike. The pods are long, thin, and held obliquely on their stalks. A native of eastern Europe and the Near East, the plant has established itself in several parts of Britain. It is common throughout the United States and is found across Canada but is most abundant in the prairie regions.

Conringia orientalis is the HARE'S-EAR CABBAGE in Britain, with erect stems 4–20in (10–50cm) tall, in the United States and Canada the HARE'S-EAR MUSTARD, growing from 10 to 40in (25–100cm). It is easily recognizable by its whitish-green rather oval leaves growing alternately and clasping the stem between two lobes, and by the pods which are 4-angled and erect on oblique stalks. The pods are 3–5 in (7.5–12.5cm) long. The flowering period is May to July. A native of Europe, this weed is a frequent casual by the sea in Britain, and is distributed throughout all the United States excepting the south-eastern and south-western areas. It extends north into southern Canada from Quebec to British Columbia.

The GARLIC MUSTARD, *Alliaria petiolata* (*officinalis*), is also known as Jack-by-the-hedge and Hedge Garlic. This is our odd-man-out among the Mustards, being a biennial and having white flowers. As its name tells us, it smells strongly of garlic. The erect stem is usually unbranched and grows 8–48in (20–120cm) tall from a basal rosette of pale green heart-shaped leaves with toothed margins. The lower stem-leaves have long stalks decreasing in length upwards. They are alternate. The flowers are borne on short stalks towards the top of the stem, ending in a cluster often surrounded by terminal leaves, developing into pods about 2in (5cm) long and nearly cylindrical, very narrow and with prominent midribs. The Garlic Mustard flowers from April to June and is usually to be seen growing in colonies in the shady part of the garden. It is distributed throughout the British Isles and Europe, locally in Quebec and Ontario and southward.

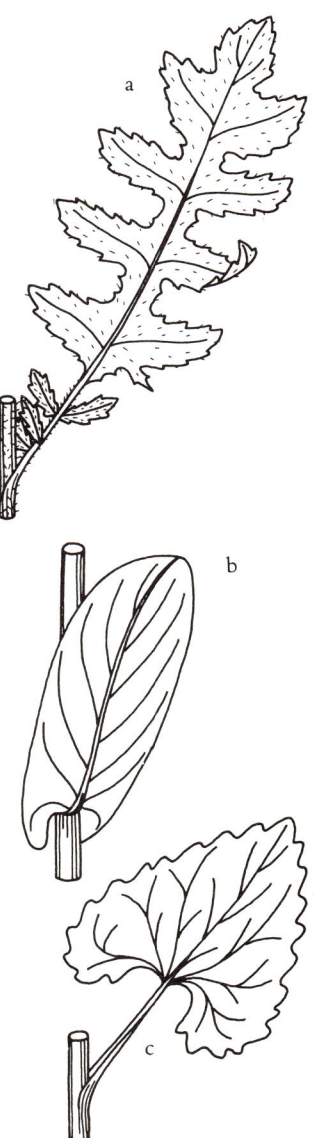

Leaf of *a*. TALL ROCKET or TUMBLE MUSTARD *b*. HARE'S-EAR CABBAGE or HARE'S-EAR MUSTARD *c*. GARLIC MUSTARD

NIGHTSHADES (Solanaceae)

BLACK NIGHTSHADE

The BLACK NIGHTSHADE, *Solanum nigrum*, carries a warning: '*I may be dangerous!*' All parts of the plant contain poisonous alkaloids but in varying amounts, so that foliage and berries (which are about the size of a pea) have been eaten by animals with no harm to them, or with fatal results. The shining yellow or black berries, attractive to children, may cause them only slight discomfort or acute suffering, exhaustion and death.

A bushy annual or biennial growing to about 2ft (60cm) tall, its leaves are alternate, pointed-oval with wavy margins, and can be translucent to thickish, dense, or opaque against the light. It belongs to the same family as the potato and has the same star-like flower, 5-petalled with 5 yellow stamens forming a prominent cone, but with branched roots instead of tubers. The flowers, white, stalked and in a loose cluster, bloom in Britain from July to September, in North America from May to November, fruiting from August to October. Found in waste places in the wild, the Black Nightshade prefers the cultivated soil of gardens and indeed is also called the Garden Nightshade. It is a European weed naturalized in North America, distributed throughout England and rarer northwards, very local in Wales, the south of Scotland and Ireland.

Another form of the Black Nightshade is *Solanum americanum*. As its name denotes, this Nightshade is a native of America and is found throughout all the eastern states and parts of the central states, and north into eastern Canada to Nova Scotia. The two are very difficult to distinguish. The unripe berries are poisonous to sheep and other grazing animals; the ripe berries are reputedly non-poisonous to humans and often eaten raw or cooked for preserves or pies.

There is no doubt about the European DEADLY NIGHTSHADE, *Atropa belladonna*, which has been a source of poison from early times. Usually 20–30 berries are needed to produce fatal results, though half a berry has been known to cause death. However, all parts are dangerous to humans, particularly the roots and seeds, the danger increasing with the plant's maturity. In animals poisoning is rare, though the flesh of rabbits that have fed on the Deadly Nightshade has, it has been stated, poisoned those who ate it. The plant is a perennial growing up to 5ft (1.5m) tall. The leaves are alternate or in unequal pairs, pointed-oval and much narrowed at the stem. The single flowers, growing from the leaf-axils, have 5 dull maroon petals and emerge from a tube-like calyx of a lurid green which is clasped at the base by 5 pointed sepals. The plant flowers from June to August. The berries, again black when ripe, are larger than those of the Black Nightshade, being about the size of a cherry. The Deadly Nightshade, also known as Dwale, is found in central and south Europe; and in England and Wales on chalk soils from Westmorland south to Somerset and Kent but only locally.

DEADLY NIGHTSHADE

Physalis heterophylla is the CLAMMY GROUND-CHERRY of the United States, the GROUND CHERRY of Canada. A native of North America, it does not occur in Europe and is not poisonous. A bushy perennial covered with sticky hairs, it has erect branching stems varying in height from 12in to 32in (30–80cm). The leaves are alternate and broadly oval, pointed at the tip and bluntly toothed. The solitary flowers are trumpet-shaped, yellow with a dark purplish centre, and hang from the leaf-axils on short drooping stalks. After flowering, the calyx inflates like a balloon, enclosing the yellow berry. Its garden cousins are *Physalis alkekengi*, the Bladder Cherry, and *Physalis franchetti*, the Cape Gooseberry or Chinese Lantern. The Ground Cherry invades gardens with gravelly and stony soils. It is found throughout eastern North America, west to Saskatchewan, Utah, central New Mexico, and west-central Texas. It is common in eastern Canada.

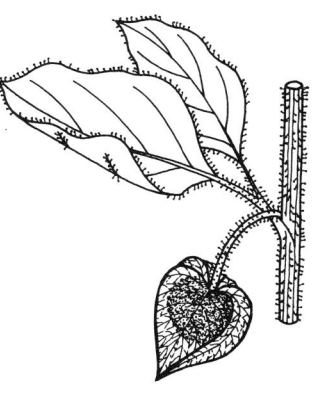

Leaf and fruit of CLAMMY GROUND-CHERRY

The American HORSE NETTLE, *Solanum carolinense*, is another non-poisonous member of the family, and recipient of a variety of names which label it undesirable: Bull Nettle (for its aggressiveness), Devil's Tomato (for its tomato-like berry which is yellow, juicy and cherry-size), Devil's Potato and Apple of Sodom. A prickly perennial with simple or branching stems, it grows 1–4ft (30–120cm) tall, reproducing not only by its seeds but by its creeping underground rhizome. The stems have star-shaped hairs with 4–8

HORSE NETTLE

rays, as well as prickles; and the prickles, rough and yellow, extend along the leaf-stalk, midrib and veins of the leaves, which are alternate, oval and coarsely toothed. The flowers are again 5-petalled but the stamens bunched instead of being in a protruding cone. The petals are violet, bluish or white, and about $\frac{3}{4}$in (2cm) across. The flowering period is May to October.

The Horse Nettle, also called the Carolina Nettle, is the host of the Colorado beetle which feeds on its leaves, and for this reason alone should be discouraged. It can be suppressed by growing taller plants that will overshadow it. Distribution is throughout all the eastern half of the United States excepting Maine; north into southern Ontario; also in the west from Oregon and California east to the Rockies in southern Idaho and Arizona.

NIPPLEWORT (Compositae)

The NIPPLEWORT, *Lapsana communis*, is recognizable by its open, rather sparse, habit, and tight green buds oval in shape. It is an annual to biennial varying in height from 1–3ft (30–90cm), with pale yellow Dandelion flowers by which you can tell the time. On a bright morning they open about 7am, and if light and temperature remain favourable they do not close till between 3 and 4pm. The florets are rather few in number, 8–15, and the 15–20 flower-heads are on slender stalks in a branched cluster, in bloom from May to September. The lower leaves are long-stalked, shaped like a lyre and pinnate in slender lobes; the upper leaves oval to lanceolate, short-stalked and sharply toothed, much smaller at the top of the stem, the texture being thin and usually hairy. All the leaves are alternate. An average-sized plant will produce nearly 1,000 fruits. A native of Europe and Asia, the Nipplewort is rare to common in the eastern United States and eastern provinces of Canada, west to the Dakotas and the west coast; common throughout the British Isles, and in continental Europe into Scandinavia. Formerly it was used as a salad plant, although its milky juice is rather bitter.

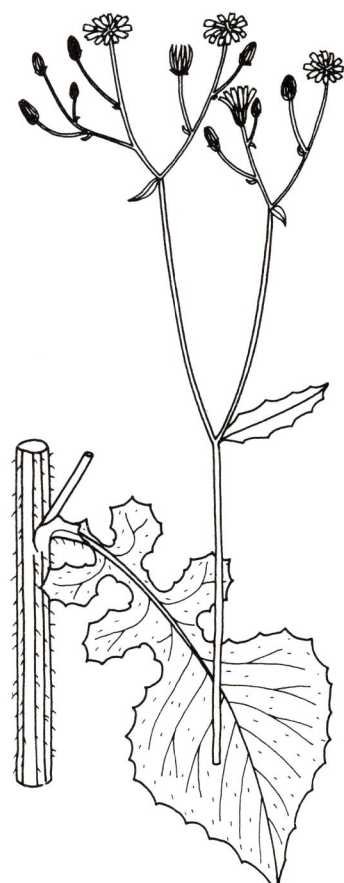

NIPPLEWORT, flowers and lower leaf

Parsleys (Umbelliferae)

The Parsleys are a handsome group of plants, seen at their best growing on waysides where they raise their heads feet above the other vegetation. I know no prettier sight than the Maytime country roads margined with the lacy leaves of 'Cow Mumble' and snowy with its flowers. Nor are the Parsleys to be deplored when they invade gardens, for their deep-foraging tap-roots bring up valuable minerals where shallower-rooted garden plants can reach them. Though don't let them seed: the Wild Carrot, for instance, produces about 4,000 seeds per plant, and 4,000 Wild Carrots would be too much of a good thing.

Cow Parsley or Wild Parsley, fruiting umbel and leaf

The Cow Parsley or Keck, *Anthriscus sylvestris*, Wild Parsley in North America, is a tap-rooted perennial 24–40in (60–100cm) tall, with wide-spreading underground stems that can soon produce a dense patch of plants. The soft bright-green leaves, alternate, up to 1ft (30cm) long and 2–3-pinnate with coarsely serrated margins, are feathery-looking and emerge from grooved sheaths on the furrowed and hollow stems which are downy below and smooth above. The inflorescence is a terminal compound umbel of tiny white 5-petalled flowers. The fruit is oblong, smooth and black, with 2 beaks at the top. The Cow Parsley is generally distributed and often extremely abundant throughout the British Isles and northern and central Europe. In North America it is found locally from

Newfoundland to Ontario, and in the eastern United States. It flowers in April and June.

FOOL'S PARSLEY, *Aethusa cynapium*, is a branched annual greatly varying in height from 2 to 50in (5–125cm) but generally about 2ft (60cm), with hollow stems bluish and finely ridged, the alternate leaves dark green and not so feathery in appearance as those of the Cow Parsley. Again the 5-petalled flowers are in compound umbels, but these are not so dense, and from the base of each hang 3–4 long bracteoles. They bloom from June to August. At the fruiting stage the stalks bend downward with the fruits erect. These are egg-shaped and ridged, with no beaks. All parts of the plant are poisonous. Fatalities have occurred when the leaves have been mistaken for garden parsley and the roots for young turnips or radishes. Though animals reject the plant because of its disagreeable smell, they accept it when contained in hay, when it is harmless. Fool's Parsley is common and generally distributed throughout the British Isles, though rarer northwards. In North America it is found locally from Nova Scotia to South Ontario and Minnesota, and southward.

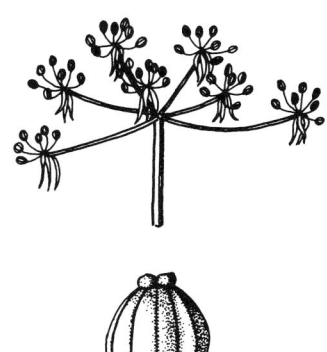

FOOL'S PARSLEY, fruiting umbel and fruit

The WILD CARROT, *Daucus carota*, more charmingly called QUEEN ANNE'S LACE, is a biennial 1–5ft (30–150cm) tall, the taller measurement applying to the United States. The stems are erect, branching, slender, hollow and ridged, bristly and hairy. It is the fine divisions into which the alternate leaves are dissected that give the plant its lacy appearance. The tiny white flowers are in dense compound umbels and have numerous bracts hanging from the base of the main umbel. The central flower is often red or purple. After flowering, the stalks of the umbel turn upward to enclose it, forming what is called a 'bird's nest'. The fruit is oblong with one side flattened, the other ridged and bristly. Flowering in Britain from June to September, in North America from May to October, the Wild Carrot is distributed throughout the British Isles except in the extreme north; in Europe from Scandinavia to the Mediterranean. In North America it is scattered across most of the United States and Canada.

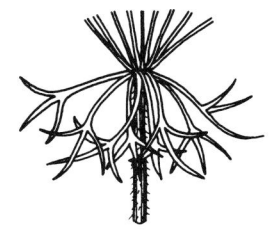

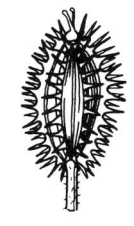

WILD CARROT or QUEEN ANNE'S LACE, bracts at base of umbel, and fruit

PEARLWORT
(Caryophyllaceae)

The PROCUMBENT PEARLWORT, *Sagina procumbens*, known in America as BIRDSEYE, has a dense rosette of linear leaves $\frac{3}{8}$—$\frac{1}{2}$in (5–12mm) long, narrowing at the tip to a bristle. From it spread prostrate stems with even smaller leaves, opposite in tufts, and from each tuft springs a long (comparatively) stalk bearing a minute white flower with 4 petals, though sometimes the petals are absent, leaving only the 4 larger green sepals. A European weed troublesome in lawns, grass verges and paths, it ranges in North America from Newfoundland to Minnesota and southward. It flowers from May to September and is automatically self-pollinating.

PROCUMBENT PEARLWORT or
BIRDSEYE

PIGWEEDS   (Amaranthaceae)

The Pigweeds belong to the same genus as the garden plants Love-lies-bleeding and Prince's Feather. Three of them are troublesome weeds widespread in North America, two of which occur in Britain as rare and impermanent casuals.

The first of these is *Amaranthus albus*, the TUMBLE PIGWEED, or TUMBLEWEED, which is another of those plants that break off at ground level when the fruits are ripe and scatter their seeds as the wind bowls them along. It is an annual, 4–39in (10–100cm), branching and with a reclining habit. The stems are whitish, the leaves alternate and broadest at the tips, tapering into a stalk at the base. The greenish flowers have sharp-pointed bracts below them and nestle in the axils in small spikes. Widespread throughout the United States and Canada the Tumble Pigweed makes a rare appearance in Britain. It flowers from July to October.

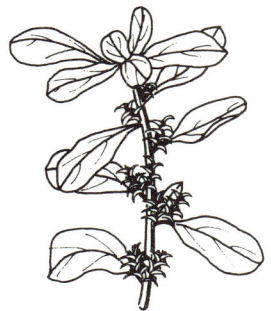

TUMBLE PIGWEED or TUMBLEWEED

The PROSTRATE PIGWEED, *Amaranthus blitoides*, has greenish or purplish annual stems radiating from the rootstock and much branched, 8in to 2ft (20–60cm) long, spreading flat over the ground and rising at the tips. The leaves are alternate on the stem but grow in bunches at the top. They are numerous, small, rounded, and tapering into the stalk. Again the flowers are inconspicuous and grow in short dense clusters in the leaf-axils: they have no petals. The bracts under the flowers are not spiny but rather soft. The flowering period is July to October. A native of the western

PROSTRATE PIGWEED

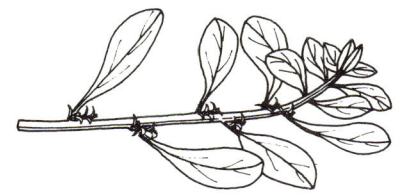

RED-ROOT PIGWEED

United States, the Prostrate Pigweed is common towards the east excepting the extreme north-east. But it is found throughout all the United States excepting the extreme south coastal areas and the extreme south-west. In Canada it has become a weed of the eastern provinces. It is not found in Europe.

The RED-ROOT PIGWEED, *Amaranthus retroflexus*, is a native of tropical America that is now practically world-wide in distribution, though a rare casual in Britain. Its stout stem, grey-green and somewhat downy, is usually branching, especially at the top: the bases of the stem may be reddish. It varies in height from 6in to 3ft (15–90cm). The drooping leaves are alternate and long-stalked, broadest near the stem: they are hairy and have prominent veins, especially beneath. Again the small green flowers have stiff sharp-pointed bracts under them, but in this species the flowers form dense spikes in the leaf-axils where they are $\frac{1}{2}$–2in (1–5cm) long, the flowering stem terminating in a cluster of these spikes crowded together and 2–8in (5–20cm) long. The flowering period is August to October. The Red-root Pigweed sometimes accumulates excess nitrates and is therefore a poison to cattle, causing them to bloat, but pigs feed on them greedily, as they do on other Pigweeds, hence the common name.

PLANTAINS   (Plantaginaceae)

Plantago major is the GREAT PLANTAIN of Britain, the
COMMON PLANTAIN of Canada and the United States,
where it is also called the BROAD-LEAF PLANTAIN, the
last name describing one of the features distinguishing it
from two of the other plantains, though not from
Plantago rugelii, the BLACK-SEED PLANTAIN. So we will
describe *major* and *rugelii* together.

Perennials, both have their oval pointed leaves
alternate in a rosette; but whereas those of *Plantago
major* are thick with minute hairs that make the leaves
rough to the touch when they are dry, the leaf-stalk
being green, the leaves of *Plantago rugelii* are thin,
smooth or slightly hairy, the leaf-stalk tinged with
purple. The flower-spikes offer the obvious difference:

GREAT or COMMON PLANTAIN

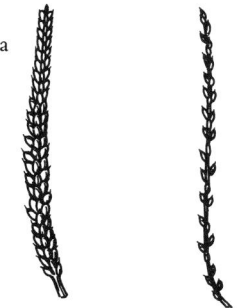

a   b

Flower-spike of *a.* GREAT
PLANTAIN *b.* BLACK-SEED
PLANTAIN

major has dense spikes 4–20in (10–50cm) long and blunt
at the end; the spikes of *rugelii* are 1ft (30cm) long,
looser, more slender, and tapering. It is likely that a
gardener in the eastern United States will know both of
these plantains, as will the Canadian gardener: *Plantago
major* is distributed the length and breadth of North
America, flowering from June to October; *Plantago
rugelii*, flowering from July to October, is confined to
the eastern side, though also found locally in British
Columbia, in Canada being called RUGEL'S PLANTAIN.
British gardeners know the Great Plantain, which is
generally distributed throughout the British Isles and
Europe, flowering from May to September.

The RIBWORT PLANTAIN of Britain, *Plantago lanceolata*, is the BUCKHORN PLANTAIN or RIBGRASS of the United States and Canada. Other names are the English Plantain, Black Plantain, and Ripplegrass. (The American name Buckhorn Plantain must not be confused with the British Buck's-horn Plantain, which is a different species, *coronopus*.)

A perennial or biennial, it is easily distinguished by its lance-shaped leaves tapering at both ends, which have prominent veins, smooth margins and lie close to the ground in a rosette or are more or less erect; and by the flower-spikes which are stubby with long stamens projecting at maturity like pins on a pin-cushion. The flowering period in Britain is April to August, in North America May to October. A European, this weed, which can be troublesome in lawns, is common throughout the British Isles, Europe as far north as Iceland, and the whole of the United States and Canada.

Flower-spike and leaf of
a. RIBWORT PLANTAIN or
RIBGRASS *b*. HOARY PLANTAIN

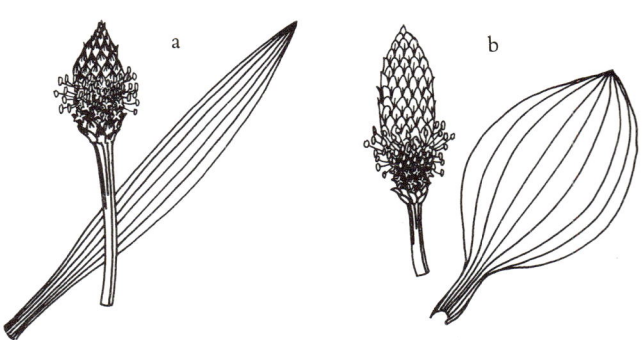

The HOARY PLANTAIN, *Plantago media*, really means hairy plantain, which applies to its broadly elliptical leaves. These, like those of the other plantains, grow in a rosette. The flower-spikes are the same stubby shape, though slightly more elongated, and the pin-cushion stamens light purple. A native of Europe, the Hoary Plantain, also called Lamb's Tongue, is rather locally distributed in the eastern half of Canada and the United States, and in British Columbia. It is fairly common in southern England and the Midlands, becoming rarer northwards; and in southern Europe. The flowering period is June to September.

POKEWEED (Phytolaccaceae)

The POKEWEED has two botanical names: *Phytolacca americana*; and in America also *Phytolacca decandra*, its synonym, meaning 10-stamened. It is a conspicuous plant ranging from 3ft to over 6ft in height (1–2m), and looks like a small tree, having a trunk-like stem with long branches which are often reddish. The generic name *Phytolacca* gives us a clue to this, the *lac* part deriving from the same source as lake, the crimson pigment. Children make red ink from the fresh berries: another name for the plant is Inkberry. The leaves are stalked, oval, and pointed at the tips, alternate on the stem. From the stem, too, grow the flowers, in opposite pairs on a spike 4–8in (10–20cm) long, curious flowers with no petals but 5 petal-like and rounded sepals, greenish-white to white or suffused with pink, cupping the pistil and the 10 stamens. The very large tap-root is poisonous, as are the seeds and leaves, though the young leaves are often cooked as greens. Boiled, the poisonous principle is poured off with the water. The phytolacca sold medicinally for cows is made from the dried roots and fruits. Pokeweed is a native of eastern North America which has been introduced into Britain and is occasionally naturalized. The flowering period is July to September.

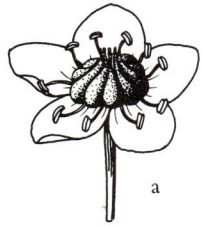
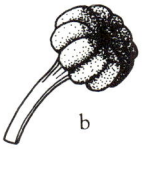

POKEWEED *a*. Flower *b*. Fruit

RED SHANK or LADY'S THUMB

POLYGONUMS
(Polygonaceae)

This group is a fantasy of pink and green — pink flowers and sometimes green, pink calyx, pink stamens, pink seeds, pink nodes, among cool green leaves. The British call many of them Persicarias, Americans the Smartweeds.

Polygonum persicaria is the RED SHANK or WILLOW WEED of the British Isles, the LADY'S THUMB of North America. It is an annual, growing 10–30in (25–75cm) tall, with stems swollen and reddish above the nodes from which grow both flower-spikes and leaves. There is a curious structure here called an ochrea, which is tubular and fringed, in colour pink. This sheath serves to protect the young flower-spike and remains on the stem. The charming little flowers, densely clustered, are 5-petalled and pink, opening from deeper pink buds. The stems of the lanceolate alternate leaves are reddish where they join at the node below the ochrea, and on the leaves are red blotches which a legend tells us are drops of blood: the plant was supposed to be growing at the foot of Christ's cross. The flowering period is June to September in Britain, June to October in North America. Common throughout the British Isles and Europe, the Red Shank is naturalized throughout Canada and the United States excepting the south-western area and Texas.

Polygonum aviculare, the KNOTGRASS or PROSTRATE KNOTWEED, is a much-branched wiry annual growing

KNOTGRASS or PROSTRATE KNOTWEED

erect when drawn up among grasses, prostrate and
radiating from the rootstock when in the open ground.
The stems are ridged with whitish ochreae, the dark
green alternate leaves varying as much as the plant's
habits, being rather linear or broadly elliptical. The
small flowers grow on spikes in clusters of 2–5 in the
leaf-axils: they have 5 sepals which look like petals,
margined with pink or white. The weed is common and
generally distributed throughout the British Isles and
southern Europe, throughout all the United States and
Canada. The flowering period in North America is July
to November, in Britain July to August but sometimes
almost the whole season.

Polygonum lapathifolium, the PALE PERSICARIA of the
British Isles, the PALE SMARTWEED of North America,
grows 1–3ft (30–90cm) tall but is often reclined, and
rooting at the basal nodes. It is a green weed, the stems
swollen above the nodes and usually greenish, the
flowers greenish-white, rarely pink, on spikes arching
or drooping. The lower ochreae are not fringed. The
alternate leaves, 2–8in (5–20cm) long, are lanceolate
and sometimes woolly beneath and often with a black
blotch at the centre. It flowers from June to September
and is generally distributed throughout Europe and the
British Isles and North America.

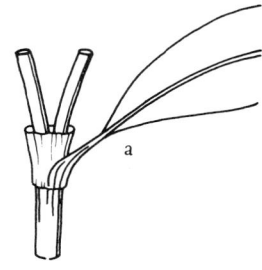

Ochrea of *a*. PALE PERSICARIA
or PALE SMARTWEED *b*. SPOTTED
PERSICARIA

Similar to the previous weed but usually smaller and
often more or less decumbent is *Polygonum nodosum*, the
SPOTTED PERSICARIA, named for the undersides of the
leaves being densely dotted with yellow glands. The
flower-spikes are pale pink, arising from pink ochreae
above thick knotted reddish nodes. The alternate leaves
are long and lanceolate. The Spotted Persicaria flowers
from July to September and is scattered throughout the
British Isles and in North America from Newfoundland
to British Columbia, south and south-westward.

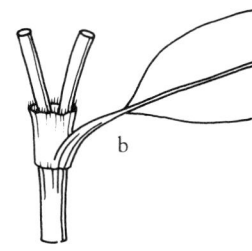

Polygonum hydropiper has many common names:
WATER-PEPPER, BITING PERSICARIA, MARSH-PEPPER
SMARTWEED, COMMON SMARTWEED. Both 'biting' and
'smart' tell of its stinging taste, to say nothing of
stinging eyes if you rub them with a hand that has
touched it. The weed is graceful in appearance, with
nodding flower-spikes of pink-tipped buds opening to
tiny greenish flowers. It grows to a height of 1–2ft
(30–60cm). The leaves are alternate and narrowly

lanceolate, smaller and narrower as they go up the stem, with blunt pink ochreae which may or may not be fringed. In Britain it flowers from July to September, in North America from June to November. *Hydropiper* means water-pepper and tells that the plant likes wet places, but it also grows in gardens wet or not. Both Britain, Europe and North America claim it as a native. It is common throughout the British Isles except in the north of Scotland; in Europe except the colder regions, and throughout Canada and the United States excepting southern Georgia and Florida.

POOR JOE (Rubiaceae)

BITING PERSICARIA or COMMON SMARTWEED

The name POOR JOE refers to the ability of *Diodia teres* to flourish in poor soil, and *Diodia* derives from a Greek word meaning thoroughfare, which tells of the waysides where it can often be seen growing. The specific name *teres* means round, thin or slender, which is a description of the leaf-like stipules at the growing-points which are fused together and form several long bristles. But don't think Poor Joe confines himself to

POOR JOE

roadsides: he will delight in good garden soil if he gets the chance. An annual with hairy branching stems arising from a shallow branching tap-root, the weed grows to a height of 4–32in (10–80cm) depending upon its environment. The spread of its branches is nearly circular. The opposite leaves clasp the stem and are narrowly lanceolate to elliptic, tapering to a long point and with smooth margins. The tiny clusters of flowers, 1–3 of them, grow in the leaf-axils: they are trumpet-like in shape and whitish-pink to lavender with 4 short lobes. This is an American weed flowering from June to October and flourishing throughout all the eastern half of the United States excepting northern New England and New York.

POPPIES (Papaveraceae)

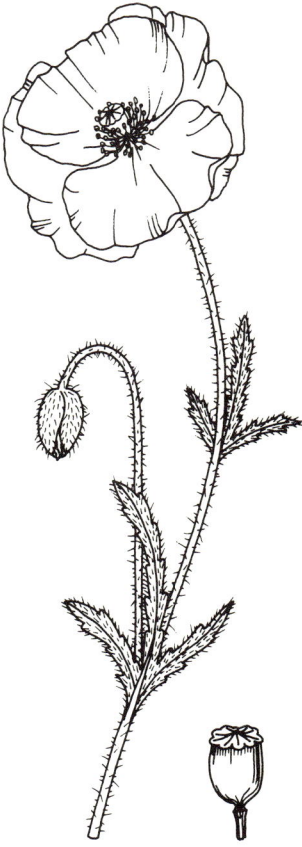

It was from the humble CORN POPPY, *Papaver rhoeas*, that the beautiful strain of garden Shirley Poppies was developed at the end of the last century by the Rev. William Wilks, the beloved and bewigged secretary of the Royal Horticultural Society. In its natural form it freely invades British and continental gardens, growing to a height of 1–2ft (30–60cm). Rarely is it seen growing alone but rather in such colonies that an entire field is crimsoned with its flowers. This can happen in your garden if you let it seed, for each plant produces about 20,000 seeds and they germinate almost on the surface of the soil. The flowers are 4-petalled with crimped edges and usually black blotches inside at the base. They grow on long stalks from the leaf-axils, the buds being hairy and nodding. The lower leaves are stalked and pinnately divided once or twice, the upper leaves clasping the stem and usually 3-lobed, the middle one being very much longer. All the leaves are alternate, the stems stiffly hairy. The Corn or Field Poppy is distributed throughout the British Isles excepting the north of Scotland; in Europe except in the far north; in North America from Nova Scotia to North Dakota, south to New England, Virginia, Missouri and Kansas. It flowers all summer.

CORN POPPY and capsule

146

Another field poppy is *Papaver dubium*, the LONG-HEADED POPPY, which describes the long seed-capsules, a feature distinguishing it from the Corn Poppy which has a cup-shaped capsule. The height is the same, the leaves similar but with shorter, broader and more abruptly acute segments, and with a smaller middle lobe. The petals always overlap at the base and are paler than those of the Corn Poppy, with no basal blotches. The seeds are bluish-black. More tolerant of poor soil, the Long-headed Poppy grows farther north in Britain; in Europe in the central and southern regions; in North America from Massachusetts to Nova Scotia, Tennessee and Missouri. The flowering period is June and July.

Capsule of *a*. LONG-HEADED POPPY *b*. OPIUM or COMMON POPPY of United States

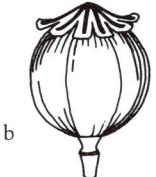

Papaver somniferum is the OPIUM POPPY, called by Asa Gray the COMMON POPPY of the United States, but still *not* common. It can become a persistent weed in gardens because one tends to tolerate it for its large white or lilac or maroon flowers which are attractive in a cloudy, somniferous kind of way, as are the undulating pale grey-green leaves, smooth and oval, lobed and irregularly toothed. The leaves are alternate, the lower ones narrowed into a short stalk, the upper ones grasping the stem. The seed-capsule is round and usually large but varying in size, the seeds black or white. This poppy, as its name tells us, is the source of the drug opium, and it contains about 20 different alkaloids of which the most important are the narcotics morphine and codeine. All parts are poisonous, especially the unripe capsules. It grows to a height of 3ft (90cm) and flowers from June to August. Cultivated commercially on a large scale in Europe, the Opium Poppy found in gardens is an escape, as it is also in the United States.

PURSLANE (Portulacaceae)

The COMMON PURSLANE or PURSLEY, *Portulaca oleracea*, is a summer weed found in all well-cultivated gardens in the United States and Canada, occasionally establishing itself for a time in Britain. Its original home is western Asia. Sprawling in habit, it has smooth, reddish, thick and fleshy stems and grows to a height of 4–12in (10–30cm). The small oval leaves, widest at the tip, are nearly stalkless and are opposite in pairs, the top ones crowded beneath the flowers which at the ends of the branches are in small clusters, solitary in the leaf-axils. They are stalkless with 5 pale-yellow petals that open only on sunny mornings. The sepals, which may be slightly notched, are white and joined to form a short tube, with the two upper ones free and pointed at the tip. The fruit is a round many-seeded capsule which splits around the middle, the upper half with the 2 free sepals on top falling away like a lid. An annual, Pursley flowers and fruits from June or July until the first frost, in hot regions from April to June, disappearing in the hottest period, reappearing in late summer and continuing until the frost. It is most abundant in the north-eastern states, least common in the Pacific north-west. In Canada it is a common weed in agricultural districts.

COMMON PURSLANE or PURSLEY

RAGWEEDS (Compositae)

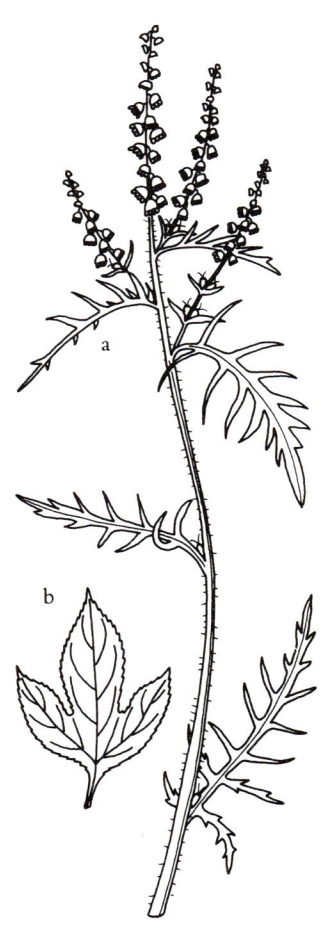

To sufferers from hay-fever *Ambrosia artemisiifolia* belies its generic name as something 'delightful to smell', for in North America the COMMON RAGWEED is one of the worst provokers of this distressing ailment when its greenish spikes of flowers fill the air with pollen. An annual, the stem is 8–36in (20–90cm) tall, erect and bluntly 4-angled, and densely covered with hairs. The leaves are alternate and long-stalked, pinnate with the leaflets lobed and alternate but with some opposite below. There are two kinds of flowers: the male flowers, which carry the pollen, growing on long-stemmed spikes from the leaf-axils; and the female flowers, fewer in number and nestling at the bases of the leaves and in the forks of the upper branches. The flowering period is July to October. A native of North America, the Common Ragweed is widespread but more abundant in eastern Canada and the eastern and north-central United States. In Britain, where it is known as the ROMAN WORMWOOD, it is an introduction locally established.

The GREAT RAGWEED or BUFFALO-WEED, *Ambrosia trifida*, is, as its name says, a much taller annual. It grows up to 10ft (3m) high, and has rough hairy stems and stalked leaves opposite in pairs. The lower leaves are 3- to 5-lobed, except those on the long branching stems bearing the flower-spikes, and these are not lobed. The flowers are similar to those of the Common Ragweed but larger. The Great or Giant Ragweed is transcontinental in distribution but more common in the western regions. In Britain it occurs as a casual.

a. COMMON RAGWEED *b.* Leaf of GREAT RAGWEED

RAGWORT (Compositae)

The RAGWORT, *Senecio jacobaea*, is TANSY RAGWORT and STINKING WILLIE in the United States, STAGGERWORT in Canada because it causes animals to stagger. It belongs to the same Daisy family as the Ragweeds, but its flowers of bright yellow ray-florets and disk-florets are much more conspicuous. They grow in clusters on long stalks at the top of the stems which range in height from 8 to 40in (20–100cm). The basal leaves form a rosette and are toothed and oval with small lobes at the base; the lower stem-leaves being stalked and pinnate, alternate on the stem and much more toothed as they grow upward. The whole plant has a ragged appearance and is the curse of neglected grazing land, strong-rooted, and poisonous to farm animals, equally a nuisance when it gets into lawns. A biennial, it flowers from July to October. It is a European and abundant throughout the British Isles. In the United States it has become newly established along the Pacific coast from Washington to northern California, and along the Atlantic coast from Maine to Rhode Island. It is common in the maritime provinces of Canada, local in Ontario, and occurs in British Columbia.

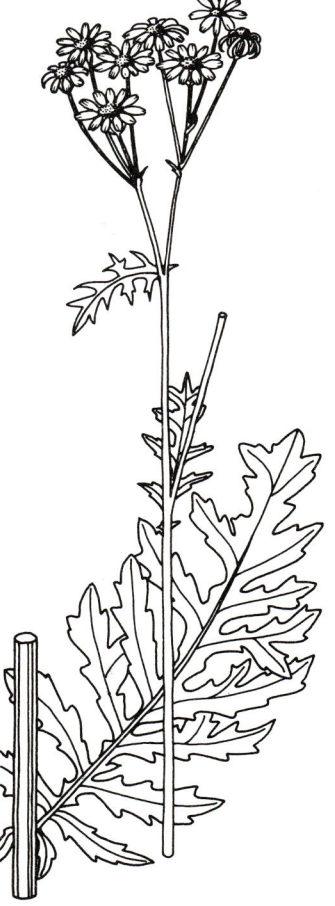

RAGWORT or TANSY RAGWORT,
flowers and stem-leaf
a. Its rosette

RED RATTLE
(Scrophulariaceae)

The RED RATTLE or SWAMP LOUSEWORT, *Pedicularis palustris*, might at first glance be confused with a pink-flowered dead-nettle, but it belongs to an entirely different family from the Labiates or lipped plants, despite the fact that it has the same top petal forming a hood, the same down-turned lower lip and a calyx that inflates after the flower has dropped off. The leaves, mainly alternate, are entirely different in shape, being delicate and feathery bipinnate, at the top of the stem much smaller than the flowers. Their stems are pink, matching the quadrangular main stems. An annual 8–18in (20–45cm) tall, the Red Rattle is much branched below, with pink and crimson flowers from May to September. This is a European weed common throughout the British Isles, in Newfoundland, the Magdalen Islands, eastern Quebec and Nova Scotia.

RED RATTLE or SWAMP
LOUSEWORT

SCARLET PIMPERNEL
(Primulaceae)

A loved little weed is the SCARLET PIMPERNEL, *Anagallis arvensis*, which has earned many names connected with its habit of closing its flower at the approach of a shower. Accordingly it is the Shepherd's Weather-glass, Poor Man's Weather-glass and, because it also closes its flower at midday, John-go-to-bed-at-noon. Another name for it is Bird's Tongue, referring to the pointed scarlet petals, of which there are 5. These pretty flowers are on long slender stalks, in pairs from the axils of the oval pointed leaves which are also in pairs, opposite, and grow spirally along the stems. An annual or perennial, it usually lies along the surface of the ground, though in favoured situations it can grow to 1ft (30cm) tall. It flowers from June to August and is distributed throughout the greater part of the world except in the tropics.

Note: If you root up this little weed, you must not fail to repeat the following charm:

> Herbe Pimpernell, I have thee found,
> Growing upon Christ Jesus' ground:
> The same guift the Lord Jesus gave unto thee,
> When He shed His blood on the tree.
> Arise up, Pimpernell, and goe with me,
> And God blesse me,
> And all that shall were thee.   Amen.

SCARLET PIMPERNEL

a

b

SMOOTH HAWK'S-BEARD   (Compositae)

The Hawk's-beards are an extremely confusing group, for the reason that 'they all look the same'. But of course each species has its difference or they *would* all be the same. They even look like the Dandelion, except that the leaves are *not* confined to a rosette, nor is the flower solitary at the end of a hollow stem.

Commonest is the SMOOTH HAWK'S-BEARD, *Crepis capillaris*, an annual weed with one or more erect stems 8–36in (20–90cm) tall. The bright yellow flowers have strap-shaped florets, the outer florets often reddish beneath, and are sometimes in pairs or single on a long stalk branching from the axils of the arrow-shaped leaves which clasp the main stem. The basal rosette and lower stem-leaves are very variable, in outline oblong to lanceolate or narrower, shaped like a lyre or pinnate with the lobes pointing backwards; the middle and upper stem-leaves are lanceolate. All the leaves are alternate but their variability extends to their surfaces: they can be smooth or somewhat hairy on one or both sides. The flowering period is from June to September. The Smooth Hawk's-beard is common throughout the British Isles and in most of Europe northwards to Denmark and the south of Sweden. It is abundant locally in North America.

SMOOTH HAWK'S-BEARD
a. Flowering stem *b*. Rosette-leaf

SORRELS (Oxalidaceae)

Several species of *Oxalis* are real pests in gardens, the bulbous ones particularly. These, introduced as garden plants, have reached infestation proportions since World War II. Weed-killers have no effect on them. Venting your rage on the bulbils by stamping on them only stimulates their growth. Chemical applications have the same effect. The bulbils will recommence growth when you give up trying. Other methods, too expensive to be practical, have been tried. Partly successful in fields overrun with *Oxalis* has been to let pigs root for the bulbils. But this does not solve a gardener's problem. He or she must wield patience and a trowel year after year if the battle is to be won, thoroughly cleaning the trowel after use, lest the merest fragment of a bulbil be carried to a new spot where it can thrive and spread. Container-grown plants and peat used for top-dressing have been blamed as carriers. Watch for seedlings, and count yourself lucky if you can gently trowel up the three-leaved invader when it has only its first parent bulb.

The WOOD SORREL, *Oxalis acetosella*, is a native of Europe and a non-bulbous perennial, represented in America by the closely allied *Oxalis montana*. It has a slender creeping rhizome with swollen, fleshy and scaly joints from each of which springs a group of single leaves in the manner of a rosette, and single flowers, both on long stalks. The bright yellow-green leaves have 3 equal heart-shaped segments which fold together

WOOD SORREL

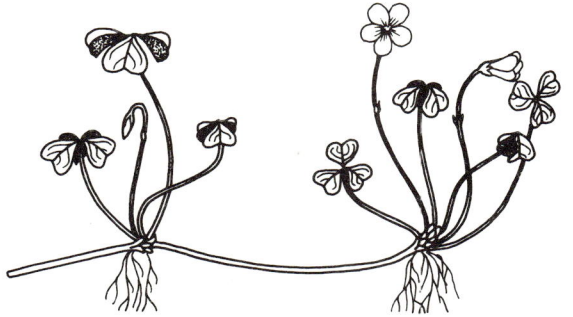

154

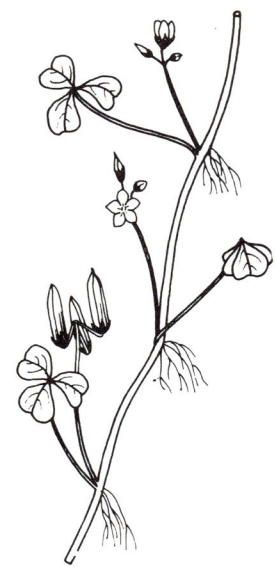

PROCUMBENT YELLOW SORREL or
CREEPING LADY'S SORREL

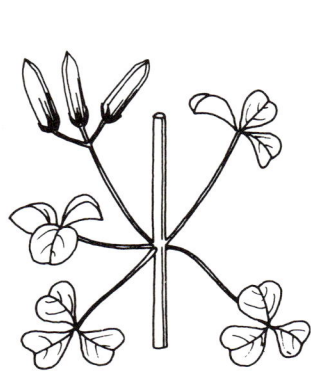

UPRIGHT YELLOW SORREL or
YELLOW WOOD SORREL

at night. The flowers, which bloom in April and May, have 5 rounded white petals veined with lilac. Although its home is in woods, this Sorrel will find its way into shady parts of the garden, and if tolerated for its undoubted charm it can soon overspread small treasured plants. It is distributed throughout Europe and Britain, and its allied species throughout the United States.

The PROCUMBENT YELLOW SORREL or CREEPING LADY'S SORREL, *Oxalis corniculata*, is another non-bulbous member of the family. Its stems come from a spindle-shaped tap-root and can be either creeping or lying loosely along the ground. It roots at the nodes and, unlike the Wood Sorrel, sends up small weak branches bearing alternate leaves, from the joints of which grow the flowers, single, in pairs or up to 7 together in an umbel on a long stalk. Again the leaves are trifoliate, the leaflets heart-shaped, often purplish, and softly hairy beneath. The flowers are yellow with 5 narrowly wedge-shaped petals making a funnel. The flowering period is May to October. The Procumbent Yellow Sorrel was first recorded in England about 1585 and is naturalized in the south. It has appeared in various places in the British Isles, and is more widely distributed throughout the tropical and warmer temperate regions of the world.

In contrast, the UPRIGHT YELLOW SORREL, *Oxalis europaea*, grows 2 to 16 inches (5–40cm) tall. It is a perennial or annual with slender underground stolons and erect stems which can be sparingly branched and either smooth or hairy. The compound leaves, again trifoliate, are whorled, tinted with purple and usually downy. The yellow flowers grow on long stalks, 2–6 in an umbel, and are in bloom from May to October. The fruit capsule is oblong. This is a troublesome weed, especially in lawns. It is common in the south of England, in Europe ranging from Scandinavia and Finland to northern Spain and Italy, but despite its specific name *europaea* it is not a native of the Old World but of the New. Until recently its synonym was *Oxalis stricta*, but this is now recognized as a separate species distinguished by the sharp angle formed by its erect seedpods and their bent stalks, *europaea* being a greener plant with straight seedpod stalks. Unfortunately in North America both are known by the same common name, YELLOW WOOD SORREL.

There is a saying in the United States that the worst weeds have come up from Mexico. This is certainly true of *Oxalis corymbosa*, one of the horrors growing from a bulb that develops into an uncountable mass of bulblets attached to the parent and copiously produced in the autumn. Leaves and flowers come from the bulb, the leaves forming a basal rosette. These are trifoliate and large, each leaflet being 1–2in (2.5–5.5cm) wide, roundish with a deep narrow groove at the apex, sparsely hairy beneath and with tiny reddish marks on the margins. The flowers have purplish-rose petals. Nobody has given this plant a common name, and the best we can do is call it the CORYMBED SORREL. The flowers are in short clusters on stalks up to 1ft (30cm) tall. Introduced into Britain, the plant has become naturalized in gardens, especially near London.

The BROAD-LEAVED SORREL, *Oxalis latifolia*, has the same habit as our previous horror but — a distinguishing feature — has its bulblets on stolons from the base of the parent bulb. The leaf-stalks are long, 4–12in (10–30cm), the leaflets ¾–1¾in (2–4.5cm), very broadly triangular with a wide shallow groove and often purplish beneath. The flowers are in umbels and are pink. This Sorrel, which is variable, is a vexatious weed in old gardens, market-gardens and glasshouses. It is a native of central and equatorial South America that has found its way across the world to the Mediterranean region, the Channel Islands and various parts of England.

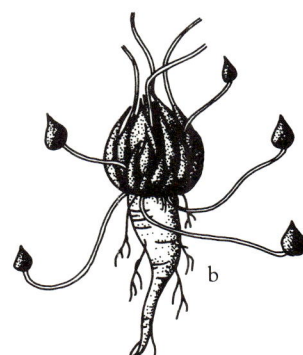

Bulb and bulbils of
a. CORYMBED SORREL
b. BROAD-LEAVED SORREL

SPEEDWELLS
(Scrophulariaceae)

The blue-eyed Speedwells are mainly annuals, that can invade a garden like a cloud of locusts. They seem to appear suddenly, already in flower, especially among other small plants where their seedlings have grown up unnoticed. The leaves are opposite, except on the flowering stems.

We will describe 10 species and divide them into two groups, 5 in each: those creeping along the ground and having solitary 4-petalled flowers; and those growing erect and having their flowers in spikes.

BUXBAUM'S SPEEDWELL or BIRD'S-EYE, *Veronica persica*, was named for the German botanist Johann Christian Buxbaum who collected plants in the Near East. It has the largest and brightest blue flower of any of the prostrate Speedwells, up to $\frac{1}{2}$in (12mm) across. The light-green leaves, $\frac{3}{4}$–$1\frac{3}{4}$in (2–4.5cm) long, are short-stalked, triangular to oval, coarsely toothed, hairy on the veins beneath. The flower-stalks are longer than the leaves, crisply hairy and springing from the leaf-axils. The seed capsule has two lobes which diverge, making it nearly twice as broad as it is long. First recorded in Britain in 1825, Buxbaum's Speedwell is now the commonest in gardens throughout the British Isles. It is distributed throughout central and southern Europe to the Mediterranean region and north to central

BUXBAUM'S SPEEDWELL or
BIRD'S-EYE

Scandinavia. In North America it is widespread but not common. It flowers from March to October.

The ROUND-LEAVED SPEEDWELL, *Veronica filiformis*, the SLENDER SPEEDWELL of North America, is tiny, its small leaves only about ¼in (5mm) across, the stalk supporting the paler blue flower no thicker than a thread though several times as long as the leaves. But, though small, it is a thoroughly aggressive weed with numerous creeping stems often forming large patches. It was introduced as a rock garden plant and soon became a troublesome escape. A perennial flowering in April and May, it spreads mainly from plant fragments and readily roots from the stems, which quickly caused its popularity to wane. Tossed on to the rubbish heap, it happily finds its way to other parts of the garden. It can run like lightning through a lawn, the mower helping it to do so. In its native Caucasian mountains it is comparatively uncommon, but it has become a serious weed in parts of Austria, France, Germany and Switzerland, and in the United States in New York and Pennsylvania. It should be burned, composted only if enough heat can be generated.

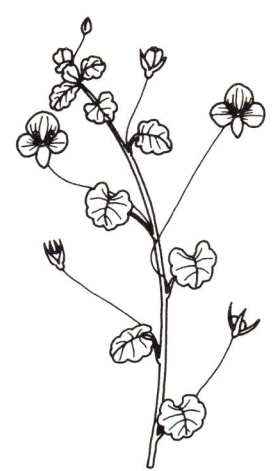

ROUND-LEAVED or SLENDER SPEEDWELL

The FIELD SPEEDWELL, *Veronica agrestis*, has leaves about the same size as *persica*, longer than they are broad, almost straight at the base and more regularly toothed. In colour they are yellow-green. The flowers are very small and usually pale blue with the lower petal or three petals white or very pale, less frequently all white, or pink on the upper surface. It flowers from February to November. The seed capsule has long hairs. This is a northerly species, a native of northern and central Europe, common in the north of the British Isles, very local in the south of England; in Europe extending into most of Norway, rare in the south. In Canada it is scattered from Newfoundland to Ontario, and has occurred in Manitoba and Alberta. It is found in the north-eastern parts of the United States.

FIELD SPEEDWELL

The GREY SPEEDWELL, *Veronica polita*, is named for its grey-green leaves which are downy, smaller than those of *agrestis*, rounder but with the same evenly toothed margins and $\frac{1}{4}-\frac{3}{4}$in (5–15mm). The flowers are small and of a deep sky-blue on long, sparsely hairy stalks, the lower petal usually whitish. They brave the February

GREY SPEEDWELL

cold and go on till November. The stems branch at the base and lie along the ground. A native of central and southern Europe, as well as western Asia and North Africa, the Grey Speedwell is found in cultivated and grassy places, particularly in vineyards and gardens, in the United States naturalized from New York to Michigan, southward and south-westward. It likes nutrient-rich, loose and somewhat sandy loams.

The last of our trailing Speedwells is the IVY or IVY-LEAFED SPEEDWELL, *Veronica hederifolia,* with stalked and rather thick light-green leaves shaped as its name declares, more or less, for the two side lobes are quite deeply notched and the centre one blunt. The upper leaves are smaller. The flower is very tiny and lilac in

IVY or IVY-LEAVED SPEEDWELL

colour, the main stems are hairy, the stalks of the flowers usually shorter than the leaves. This Speedwell is distributed throughout the whole of Europe including the British Isles, western Asia and the Mediterranean region. It is naturalized in North America. In gardens, look for it beneath bushes. It flowers from March to May.

Now we come to the speedwells that grow erect and carry their flowers on spikes. The COMMON SPEEDWELL, *Veronica officinalis*, while a perennial plant with stems creeping and rooting, often forming large mats, sends up erect spikes of long-stalked flowers, pale blue to lilac or lavender. These grow from the axils of the leaves which are in pairs opposite each other, oval, toothed along the margins and hairy on both sides, $\frac{3}{4}$–$1\frac{1}{4}$in (2–3cm) long. The flowering period is May to July. Both Britain and the United States claim it as a native, though it is also native to Eurasia. It is common throughout the British Isles, in continental Europe from Iceland and Scandinavia to central Spain and Portugal, Sardinia, Sicily and northern Greece. It is distributed throughout the eastern half of the United States excepting the extreme south, and north into Canada from Newfoundland to Ontario.

COMMON SPEEDWELL

Veronica arvensis is the WALL SPEEDWELL of Britain, the CORN SPEEDWELL of the United States and Canada. It is fibrous-rooted with soft hairy stems, single or branching, erect or nearly so, growing to a height of 2–16in (5–40cm). The leaves are triangular to oval, the lowest stalked and in opposite pairs; the upper ones alternate and growing close to the stem. The inflorescence occupies about two-thirds of the height of the plant, small pale blue flowers in the leaf-axils ending in a dense spike at the top. An annual, it is in flower from March to September. This is a weed that troubles both lawns and flower borders. It is common throughout the British Isles and Europe; throughout the eastern half of the United States, along the Pacific coast eastward to western Montana, eastern Colorado and south-eastern California; north into Canada from Newfoundland to British Columbia.

WALL or CORN SPEEDWELL

PURSLANE or AMERICAN
SPEEDWELL

Veronica peregrina can be an annual or winter annual. In the United States and Canada it is commonly called the PURSLANE SPEEDWELL, and because it is a native of America it is called in Britain the AMERICAN SPEEDWELL. The branching stems 2–16in (5–40cm) tall grow from a fibrous root and are erect, spreading and branching, smooth or hairy and rather fleshy. The lower leaves are elliptical on long stalks from the base; the stem-leaves are opposite and smaller and narrower upward. Like the Wall or Corn Speedwell the Purslane Speedwell carries the flowers in the leaf-axils but all the way to the top; that is, without their becoming a dense spike. In North America it flowers from March to August, in Britain from April to July. This is another lawn weed with a liking also for borders. It is distributed throughout all the eastern half of the United States excepting northern Maine; across the northern border, right across Canada, and south to central California, with a distinct area in Colorado. Introduced into Britain, it is found mainly in Ireland and a few places in England. It is also naturalized in western and central Europe.

The THYME-LEAVED SPEEDWELL, *Veronica serpyllifolia*, is a perennial 4–12in (10–30cm) tall with softly hairy stems creeping and rooting at the nodes. The leaves are light green, oval and smooth and opposite on the stems, continuing into a spike 4in (10cm) long or longer of up to 30 small bluish-white flowers, each in the axil of a bract-like leaf. The petals have dark blue lines pointing inward to the nectary, the honey-guides for the insects that pollinate them. The Thyme-leaved Speedwell is a native of Europe and is distributed from Iceland and Scandinavia to central Spain and Portugal, Sicily, Albania and Greece, across North America and throughout Britain. It flowers from March to October.

THYME-LEAVED SPEEDWELL

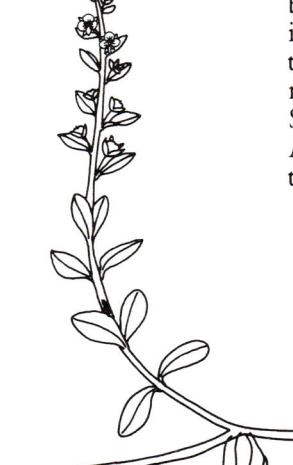

Veronica chamaedrys, the GERMANDER SPEEDWELL of Britain, confusingly in North America another BIRD'S-EYE, is the largest-flowered species of our erect and spiked Speedwells, and in Britain the tallest. A perennial growing to a height of 8–16in (20–40cm), the stems are first prostrate and rooting at the nodes but thereafter erect. They have long white hairs in two lines on opposite sides and are smooth between them. The dull-green hairy leaves are opposite, oval with round-toothed margins, and are either short-spiked or growing close to the stem. The flower spikes are borne on long stalks from their axils in a lax spike of 10–20 flowers, deep blue with a white eye, each on a short stalk from an axil. The flowering period is March to August. A native of Eurasia, the Germander Speedwell or Bird's-eye is very common throughout the British Isles, in Europe from Scandinavia to central Spain and Portugal, northern Italy and Greece. It is naturalized in North America.

GERMANDER SPEEDWELL or BIRD'S-EYE

SPURGES (Euphorbiaceae)

The Spurges are known for their acrid milky juice that can cause considerable irritation to weeding fingers, face and eyes. Another distinguishing feature of most species is the yellowy-green colour of their leaves and the petal-like bracts round the flowers. They are not ordinary flowers but naked ones, having no petals. They sit in a group, a single female (consisting only of an ovary on a stalk which elongates when it is mature and ready to scatter its seeds) surrounded by a number of males (single stamens on jointed stalks). The botanical name comes from Euphorbus, physician to King Juba II of Mauretania who named a North African species in his honour. The common name is a corruption of 'purge', for which purpose the early physicians used it, with drastic and often fatal results.

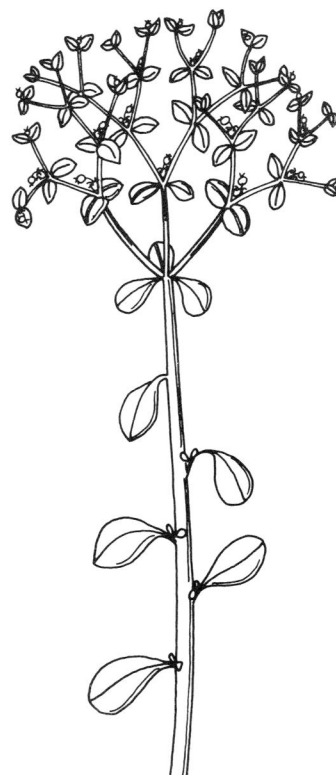

Commonest in British gardens is the PETTY SPURGE, *Euphorbia peplus*, an annual 4–12in (10–30cm) tall with a stem branched from ground-level upward. The leaves are short-stalked, smooth and oval, broadening towards the tip. The flowers grow in a 3–rayed branching umbel with a whorl of 3 leaves beneath; and beneath each pair of branches is a pair of leaves. The Petty Spurge flowers from May to November, and is also common all over Europe, in most Canadian provinces and the northern United States. Each plant yields about 1,200 seeds.

PETTY SPURGE

A taller annual weed is the SUN SPURGE, *Euphorbia helioscopia*, a native of the Mediterranean and central Asia common throughout the British Isles and found across most of North America where it is called WARTWEED. Besides its greater height, 4–20in (10–50cm), it is recognizable by its leaves having

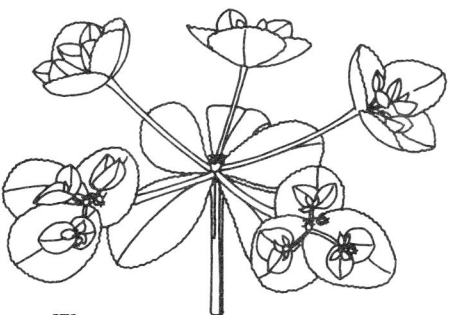

SUN SPURGE or WARTWEED

slightly toothed margins. The leaves are again broadly oval, like a club, alternate on the stem. The umbels have 5 rays, beneath which is a whorl of 5 leaves; and at the end of each ray, beneath the group of flowers, is a whorl of 3 leaves. The flowering period is April to November.

Taller again is the biennial CAPER SPURGE, *Euphorbia lathyrus*: that is, in its second year when it is mature and produces flowers, when it attains a height of up to 3ft (90cm); in its first year it forms only a short erect leafy stem. It started its garden career as an ornamental plant from continental Europe, attractive with its purplish-red stems and dark green leaves sometimes tinged with purple, useful too as well as decorative with its round green fruits often pickled as a substitute for capers, and nicknamed the Mole Plant, because moles are said never to disturb the ground where it grows. The leaves are in opposite pairs clasping the stem and growing spirally up it, so that each two pairs point in four directions. They are 3–4in (7.5–10cm) long and are narrowly oblong. The umbels are 2–6-rayed, once or twice forked, the bracts beneath them bright yellow-green and again clasping the stem. The Caper Spurge flowers in June and July. Being originally a garden plant, it is more commonly found in gardens than in the wild, but it occurs in woods in a few places in England and Wales, and in Scotland north to Lanark and Fife; in North America in southern New England to Ohio and south to North Carolina. In Europe it is distributed from Spain, northern Italy and Greece northward to France and Germany.

The SPOTTED SPURGE or NODDING SPURGE, *Euphorbia nutans* (*maculata*), is a native of the United States and not found in continental Europe or the British Isles. It has conspicuous reddish blots towards the base of the leaves which are irregularly oval, slightly toothed, and in opposite pairs on stalks with thickened joints. This is an annual with a shallow tap-root, a stem unbranched or much-branched, erect or spreading. It grows to a height of over 3ft (1m) and flowers from June to October. The

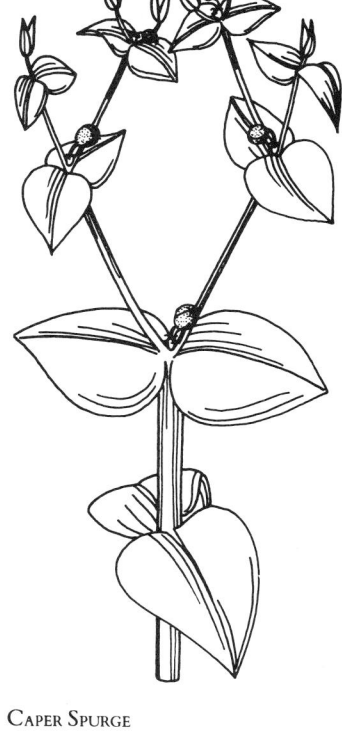

CAPER SPURGE

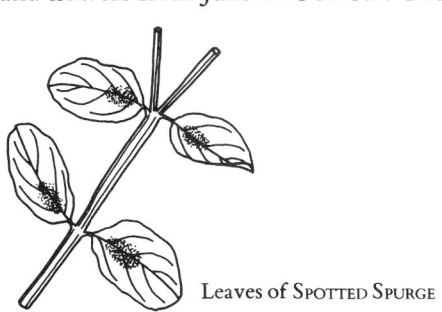

Leaves of SPOTTED SPURGE

flowers are solitary or clustered with minute petals in the form of a cup, the floral stems emerging from the leaf-axils and also bearing pairs of opposite leaves. The Spotted Spurge is distributed throughout the eastern two-thirds of the United States, and along the Pacific coast from Washington to central California.

Another spotted member of the genus is *Euphorbia maculata* (*supina*), the PROSTRATE SPURGE, much smaller than *nutans* and lying flat on the ground, but with the same blots on its leaves. This too is an American weed not found in continental Europe or the British Isles.

We include two European Spurges that belong to the same family but to a different genus. These have a watery juice instead of latex. The first is DOG'S MERCURY, *Mercurialis perennis*, a hairy perennial with long creeping rhizomes throwing up single erect stems to a height of 6–16in (15–40cm). The long oval and pointed leaves have toothed margins and are opposite in pairs, each two pairs as in the Caper Spurge pointing in four directions. Below each pair are 4 tiny bracts. The floral spikes grow from leaf-axils — the male flowers on long slender spikes, the female flowers on shorter spikes. All parts of this plant are very poisonous, especially the young shoots in early spring. Dog's Mercury flowers from February to May and is common throughout most of Britain, in continental Europe ranging from Scandinavia to central Spain and Portugal, Corsica, Sicily and Greece.

DOG'S MERCURY *a*. Female plant *b*. Male plant

The ANNUAL MERCURY, *Mercurialis annua*, is not so common as the perennial species but is equally poisonous. It has an erect smooth stem with opposite branches and grows 1ft (30cm) tall, flowering from July to October. The oval-lanceolate toothed leaves which are also opposite are a dull dark shining green with a foetid smell. The flowers are green, the male ones on long axillary spikes, the females in the axils, usually 2 together. The Annual Mercury, originally introduced as a medicinal plant, is more common in gardens and around towns and villages than in the wild. It is widespread but local in southern England but extends to Lancashire and Northumberland, in Scotland in the east, in Ireland in the south and north to Dublin and Clare. It is spread throughout most of continental Europe, and locally from Quebec to Ohio and southward as BOYS-AND-GIRLS.

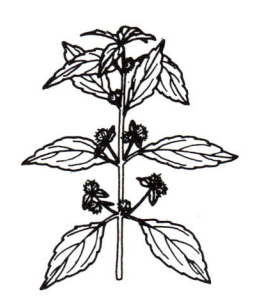

ANNUAL MERCURY or BOYS-AND-GIRLS, female plant

SPURREY   (Caryophyllaceae)

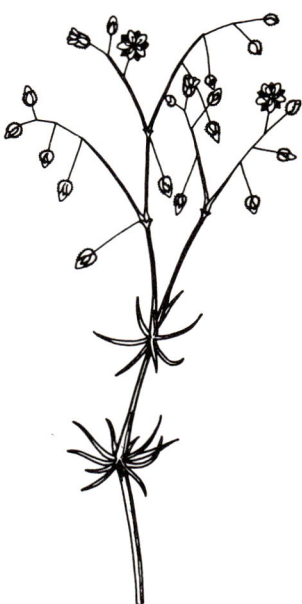

In appearance, the CORN SPURREY, *Spergula arvensis,* is similar to *Galium aparine*, the Goosegrass or Catchweed Bedstraw which we have already described. It has the same spreading kind of growth, the same sticky stems and whorled leaves. But where the leaves of the Goosegrass are lanceolate, those of the Corn Spurrey are linear. The arrangement of the flowers is different: those of the Bedstraw grow in umbels from the leaf-axils all along the stem; those of the Corn Spurrey are terminal and in open clusters. The Goosegrass flower has 4 white petals, the Corn Spurrey 5 white petals with a flowering period from March to October. An annual, it grows 6–18in (15–45cm) high. A locally abundant weed, it is spread throughout the British Isles and continental Europe, and is locally common in the eastern United States and provinces of Canada.

CORN SPURREY

STINGING NETTLES
(Urticaceae)

The big STINGING NETTLE, *Urtica dioica*, is a weed that is valuable, its coarse ugliness hiding the wealth of good uses to which it can be put. Read all about its virtues in chapter II of this book. A perennial, it can range in height from 1 to 5ft (30–150cm), and is usually seen growing in a clump, often where a bonfire has been or where rubbish has been dumped in a neglected corner. It is hairy all over with dark green leaves sharply toothed, the roots being much branched, tough and yellow, with

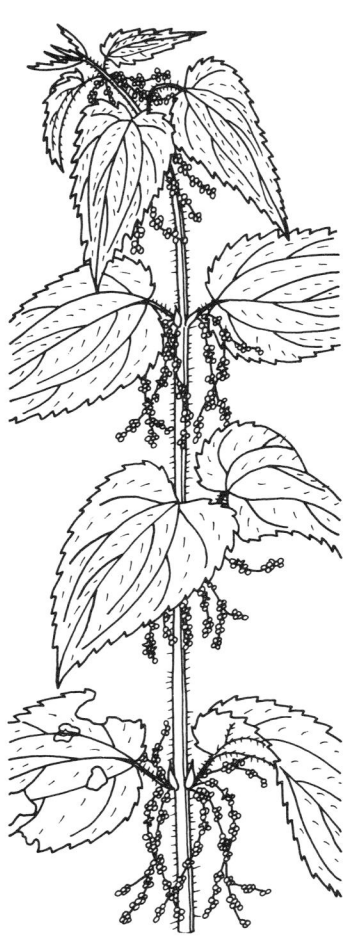

stems creeping and rooting at the nodes. The oval pointed leaves grow in opposite pairs, each two pairs forming a cross, for in this way the leaves can catch all the available light to store the chlorophyll in which the plant is so rich. The tiny green female flowers hang in dense catkins from the leaf-axils, the male flower-spikes ascending. It is the hairs on the leaves and stem which produce stings when you touch them, each hair having a sac containing the irritant fluid and bearing a stiff tapering tube. The merest touch will break the brittle tip of the tube which then enters the skin like a hypodermic syringe and injects the poison. We are advised to 'grasp' a nettle rather than touch it gently, because in grasping it the hairs are not broken off at the tips and therefore do not penetrate. Flowering from June to September, the Stinging Nettle is common throughout the British Isles and continental Europe, throughout all the United States excepting southern Georgia, most of Florida and the area from north-western Washington through most of Texas.

Though the annual SMALL NETTLE, *Urtica urens*, the BURNING or DOG NETTLE of the United States, can attain a height of 2ft (60cm) it is often noticed scattered in flower borders as a youngster of only a few inches. But it can still sting painfully, though occasionally it is devoid of the stinging hairs. The leaves are placed in the same way as those of its larger cousin but the veining is different, not branched but running in lines to the tip. Again the inflorescences spring from the leaf-axils, but they are in short horizontal spikes. The Small Nettle flowers from June to September and is distributed

STINGING NETTLE, female plant

SMALL or BURNING NETTLE, flowers and leaves

locally throughout the British Isles, across Canada to British Columbia, south to Nova Scotia, New England, New York, Pennsylvania, West Virginia, Illinois, Missouri and California.

THISTLES (Compositae)

Cirsium arvense, called in Britain the CREEPING THISTLE, on the other side of the Atlantic the CANADA THISTLE, is America's worst weed, its slender tap-root producing far-creeping whitish lateral roots that can develop an immense colony in a few years, and this from a single seed. But even a small bit of the root can start a colony, sending up spiny-edged leaves as it develops its own root-system, so that mature and half-grown and new shoots are all to be seen growing in the one patch. A perennial averaging 2ft (60cm) in height, this thistle belongs to the Daisy family, and so the flower-head is composed of florets, and these can be a dull pale purple, whitish, or pink, bunched in a flat spreading tuft above the involucre, a calyx-like structure vase-shaped and made up of overlapping purple bracts. The flower-heads are short-stalked, solitary from the leaf-axils or in terminal clusters of 2–4. The basal leaves are oblong to lanceolate in outline narrowing to a short stalk, with twisted lobes ending in strong spines; the middle and upper leaves similar but clasping the stem, more deeply cut, all the leaves being alternate, smooth on both sides or cottony beneath. The flowers are strongly honey-scented and visited by many kinds of insects. The flowering period is July to October. Despite its common name in America, this thistle is not native to Canada though as aggressive there as in the United States where it is outlawed in every northern state. In all, thirty-seven states have legislated against it. It is an equally troublesome and abundant weed throughout the British Isles and in continental Europe from Scandinavia to the south.

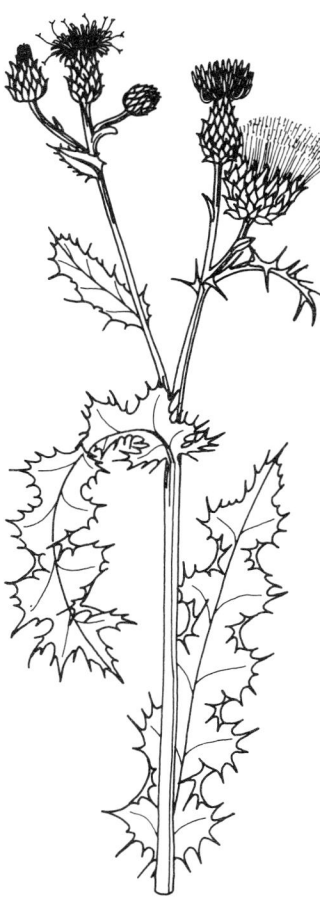

CREEPING or CANADA THISTLE

Cirsium vulgare is Britain's SPEAR THISTLE and America's BULL THISTLE, a biennial reproducing only by seed. It has a large fleshy tap-root and varies in height from 1 to 5ft (30–150cm). First-year plants form only a rosette of leaves, oblong to lanceolate or elliptic and coarsely toothed. In the second year the stems grow erect: they are furrowed, cottony-hairy and have spiny wings at the base of the alternate leaves which are lobed and spiny with long needles at the tips, the whole surface rough and spiny, woolly or merely hairy

Head of SPEAR or BULL THISTLE

beneath. The flower-heads are compact, the deep-purple or rosy florets emerging from a round or vase-shaped spiny involucre green in colour. It flowers from June to October and is an aggressive weed, though more easily dealt with than the creeping-rooted Canada Thistle. It is common throughout the British Isles, continental Europe from Scandinavia to the south, the United States; and common to infrequent in Canada.

Three common thistles belonging to a different genus all have yellow flowers. They are not nearly so prickly, and when broken they exude a milky latex.

The FIELD MILK-THISTLE, *Sonchus arvensis*, is the PERENNIAL SOW-THISTLE of the United States, the SOW THISTLE of Canada. It has creeping underground stems from which it grows erect to a height of 2–5ft (60–150cm). The stems are furrowed, hollow and very hairy. The alternate leaves are crowded on the lower part of the stem, deeply lobed, oblong to lanceolate and narrowing into a winged stalk; the upper leaves scarce, often without lobes, and clasping the stem. The flower-heads are clustered at the top, the strap-shaped florets a bright orange-yellow, the branches bearing them densely covered with yellow hairs, as are the green ribbed involucres. In Britain it flowers in August and September, in the United States June to October, and as early as April in the warmer regions. It is locally frequent to occasional throughout the northern United States, rarer in the southern, central and south-western areas, common across Canada and throughout the British Isles and continental Europe.

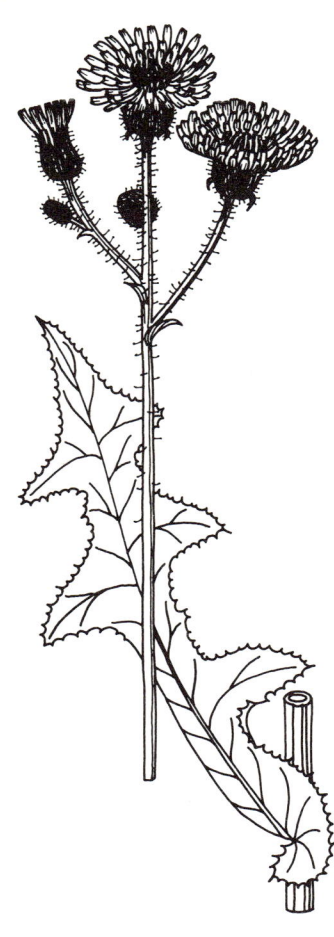

Flowering stem and stem-leaf of FIELD MILK- or PERENNIAL SOW-THISTLE

A prickly member of the genus is *Sonchus asper*, in Britain called the SPINY MILK- or SOW-THISTLE, in the United States the SPINY-LEAVED SOW-THISTLE, in Canada the SPINY ANNUAL SOW-THISTLE. It grows from a stout tap-root to an average height of 2ft (60cm), the stems smooth and often reddish, the alternate leaves again crowding along the stem, especially on the lower part where they are also more deeply divided. All the leaves clasp the stem, the two basal lobes protruding like ears. The flowers are golden-yellow, the ribbed involucre pear-shaped or round. The long stalks of the clustered flower-heads emerge from the top of the stem at an encircling spiny leaf. The flowering period is from June to August; in the southern states of Florida, Texas and California throughout the year. This thistle is common all over the British Isles, in Europe north into Scandinavia, and throughout the whole of the United States and Canada.

SPINY MILK- or SPINY-LEAVED SOW-THISTLE

Sonchus oleraceus is Britain's MILK- or SOW-THISTLE, in North America the ANNUAL or COMMON SOW-THISTLE. It has a long, slender, pale tap-root and stout, erect, smooth stems growing to a height of 8–60in (20–150cm). The stems are 5-angled, hollow except at the joints, and branched above. The main leaves are distinctively lobed, 2–3 each side of the midrib, with a longer, broader lobe at the top. The upper leaves may be unlobed. All are alternate, the margins with weak prickly teeth. The flower-heads, emerging from the axil of a leaf lobed at the base, are clustered at the top, the flowers pale yellow, the involucre again ribbed but blunt at both ends. The flowering period is June to September. This thistle is common throughout the British Isles, continental Europe and North America.

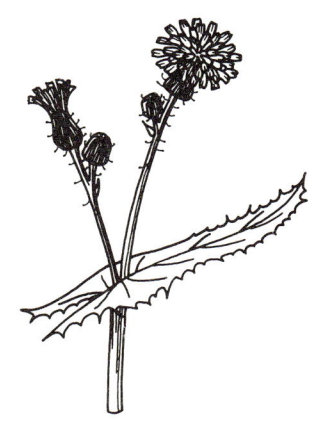

MILK- or ANNUAL SOW-THISTLE

VERVAIN (Verbenaceae)

The PROSTRATE VERVAIN, *Verbena bracteata*, is an American weed spread throughout most of the United States and Canada. Despite its common name few of its stems lie along the ground, and its erect branches attain a height of up to 16in (40cm). It is a very stemmy weed, for its leaves are sparse. These are oval in outline, coarsely toothed and with two obtuse-angled lobes at the stalk. They are roughly hairy and grow in opposite pairs, with a fresh pair emerging from the axils. The purple-blue tubular flowers spread at the top into 5 petals. They grow in dense terminal spikes, almost hidden by stiff hairy bracts. An annual, it flowers from June to August and is an invader of lawns.

PROSTRATE VERVAIN with detail of flower

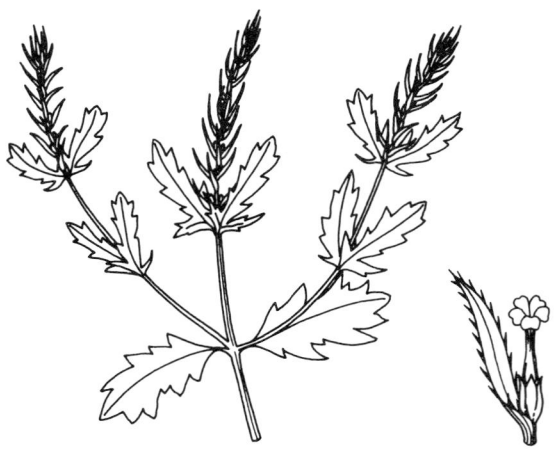

VETCHES (Papilionaceae)

The COMMON VETCH, *Vicia sativa*, the TARE or SPRING VETCH in North America, is a trailing or climbing annual reaching up or along 6in to 4ft (15–120cm). Its reddish or bluish-purple pea flowers grow in the leaf-axils, in pairs or singly. The stipules below the flower are toothed and usually marked by a dark spot in the centre. The beaked pods are smooth, 1–2in (2.5–5cm) long and rather narrow, with 10 or 12 smooth round seeds. The flowering period is May to July. *Sativa* means 'cultivated': the Common Vetch has long been used as a fodder plant and is widely distributed in temperate regions. The leaves are pinnate with 4 to 7 pairs of leaflets oval with ends either blunt or slightly notched, and instead of the top leaflet there is a 3-forked tendril by which the plant hoists itself up or over obstacles.

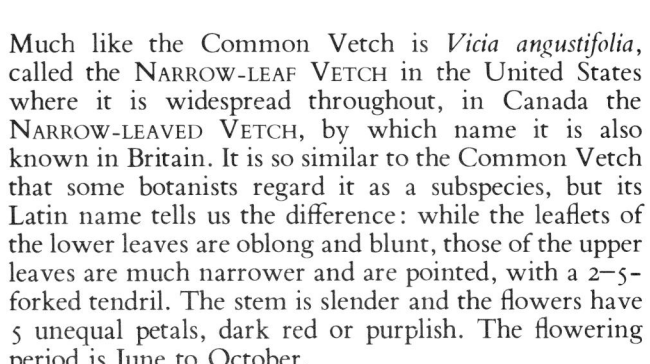

Much like the Common Vetch is *Vicia angustifolia*, called the NARROW-LEAF VETCH in the United States where it is widespread throughout, in Canada the NARROW-LEAVED VETCH, by which name it is also known in Britain. It is so similar to the Common Vetch that some botanists regard it as a subspecies, but its Latin name tells us the difference: while the leaflets of the lower leaves are oblong and blunt, those of the upper leaves are much narrower and are pointed, with a 2–5-forked tendril. The stem is slender and the flowers have 5 unequal petals, dark red or purplish. The flowering period is June to October.

COMMON VETCH or TARE
a. Leaf of NARROW-LEAVED or NARROW-LEAF VETCH

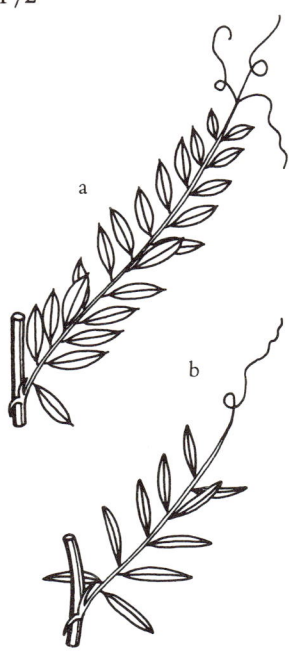

Leaf and tendril of *a*. TUFTED VETCH *b*. SMOOTH TARE or FOUR-SEEDED VETCH

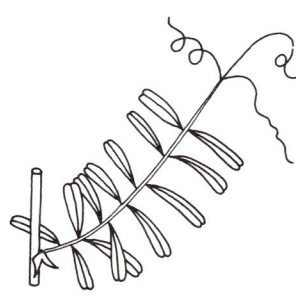

Leaf and tendril of HAIRY TARE

The TUFTED VETCH, *Vicia cracca*, is a climbing perennial with dense spikes of beautiful blue-purple flowers arising on long stalks from the axils of the leaves which are alternate and pinnate, having 6–12 or more pairs of very narrow leaflets, lanceolate on the lower leaves, linear on the upper ones and having a 2–5-forked tendril. It flowers from June to August and can climb as high as 6ft (1.9m). It is generally distributed and common throughout the British Isles, continental Europe, the United States, and in Canada from Newfoundland to British Columbia.

Vicia tetrasperma means the FOUR-SEEDED VETCH, and this is the name it goes by in North America. In Britain it is called the SMOOTH TARE. It is a slender little annual flowering from May to August, which, although endowed with a tendril at the end of the leaf, prefers to scramble over the ground where it often forms dense mats. The pairs of pale blue flowers are each stalked at the end of a long common stem from the leaf-axils. The leaves are again alternate and pinnate, with 4–6 pairs of linear leaflets and a single tendril. This weed is distributed throughout England, Wales and southern Scotland, local in the north of Scotland, naturalized in a few places in Ireland; in Europe as far as the southern half of Scandinavia; in North America local to common in the east and on the west coast.

The HAIRY TARE, *Vicia hirsuta*, is another slender, trailing annual but with small whitish or very pale mauve flowers in clusters, occasionally a single flower but up to 9. The downy pod has only 2 seeds, the long-stalked alternate leaves 4–8 or even 10 pairs of leaflets linear to oblong with a 2–4-forked tendril. As a weed of cultivation the Hairy Tare flowers from May to August and is common throughout the British Isles and continental Europe, in North America from Newfoundland to British Columbia and southward.

WHITLOW GRASS
(Cruciferae)

The WHITLOW GRASS, *Erophila verna*, in North America *Draba verna*, is not a grass at all but a tiny white-flowered rosette-leaved nuisance that appears in lawns, paths and borders, rockeries and even in walls, and starts flowering in February before the weeding season has begun. Its maximum height is 8in (20cm), but at $\frac{3}{4}$in (2cm) it is hardly noticeable. It goes on flowering till May. Being a Crucifer it has 4 petals but so deeply lobed they look like 8. The leaves are spatulate to lanceolate, hairy and with or without toothed margins. From the centre of the rosette rise several long stems each bearing an inflorescence of minuscule flowers on alternate stalks, the lower ones podding while the topmost is still in bud. A native of Eurasia, it is common and widely distributed throughout the British Isles and continental Europe, in the United States from Massachusetts to Illinois, Kentucky, Tennessee and North Carolina.

WHITLOW GRASS

WILLOW HERBS
(Onagraceae)

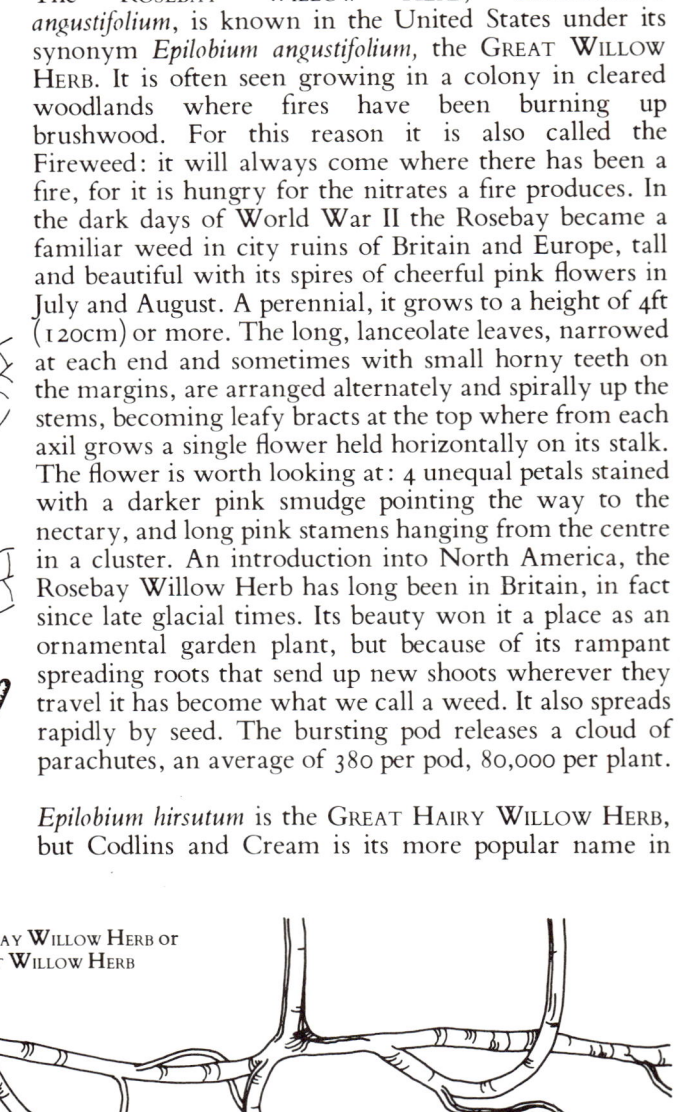

The ROSEBAY WILLOW HERB, *Chamaenerion angustifolium*, is known in the United States under its synonym *Epilobium angustifolium,* the GREAT WILLOW HERB. It is often seen growing in a colony in cleared woodlands where fires have been burning up brushwood. For this reason it is also called the Fireweed: it will always come where there has been a fire, for it is hungry for the nitrates a fire produces. In the dark days of World War II the Rosebay became a familiar weed in city ruins of Britain and Europe, tall and beautiful with its spires of cheerful pink flowers in July and August. A perennial, it grows to a height of 4ft (120cm) or more. The long, lanceolate leaves, narrowed at each end and sometimes with small horny teeth on the margins, are arranged alternately and spirally up the stems, becoming leafy bracts at the top where from each axil grows a single flower held horizontally on its stalk. The flower is worth looking at: 4 unequal petals stained with a darker pink smudge pointing the way to the nectary, and long pink stamens hanging from the centre in a cluster. An introduction into North America, the Rosebay Willow Herb has long been in Britain, in fact since late glacial times. Its beauty won it a place as an ornamental garden plant, but because of its rampant spreading roots that send up new shoots wherever they travel it has become what we call a weed. It also spreads rapidly by seed. The bursting pod releases a cloud of parachutes, an average of 380 per pod, 80,000 per plant.

Epilobium hirsutum is the GREAT HAIRY WILLOW HERB, but Codlins and Cream is its more popular name in

ROSEBAY WILLOW HERB or
GREAT WILLOW HERB

Britain, deriving from the Codling apple of long tapering shape, which is the shape of the funnel flowers, the leaves tasting like the apple, the white stamens and stigma adding the 'cream'. Great the plant certainly is, the tallest of the Willow Herbs, often attaining a height of 5ft (150cm). Hairy it is, too; the stems are densely downy and with numerous spreading hairs; the leaves, alternate, lanceolate and clasping the stem, are hairy on both sides and especially on the veins. Not only is it a taller plant — the showy purplish-pink flowers also are larger, the 4 petals notched. The Great Hairy Willow Herb is a perennial flowering from June to August, spreading by its white fleshy underground stems and by its multitude of seeds. It is found throughout Great Britain northwards to Perth and Angus, and north-east Ireland; in Europe northwards to the south of Sweden. It is locally common in south-eastern Canada, the southern parts of New England, and parts of the Mid-west.

Epilobium parviflorum, the European SMALL-FLOWERED HAIRY WILLOW HERB, is smaller in every part, the leaves narrower, but it still grows to a height of 1–3ft (30–90cm). There are several differences between the two: instead of creeping underground stems, the Small-flowered Hairy Willow Herb produces above-ground leafy stolons, and the hairs on the stems are soft and short. Its flowers are a paler pink and the 4 lobes of the stigma spread but do not turn under as they do with the stigma of its bigger cousin. Also, the stigma is about the same height as the stamens, whereas that of the Great Hairy Willow Herb is taller than the longest stamens. In its native haunts the Small-flowered Willow Herb likes wet places. It is common throughout the British Isles, and in Europe northward as far as the south of Sweden.

Another European, the BROAD-LEAVED WILLOW HERB, *Epilobium montanum*, might be called Codlins and Cream's poor relation, for its narrower funnel flowers are a washed-out pink and insignificant. They grow singly and also in a cluster at the top of the often reddish stem and have 4 petals deeply notched. The leaves, which are mostly opposite, sometimes in whorls of 3, are short-stalked, oval and toothed. This Willow Herb is less tall, 8–24in (20–60cm). It flowers from June to August throughout continental Europe and Britain.

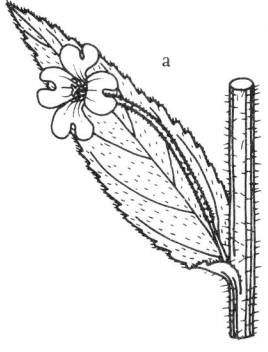

Flower and leaf of *a*. GREAT HAIRY WILLOW HERB *b*. SMALL-FLOWERED HAIRY WILLOW HERB

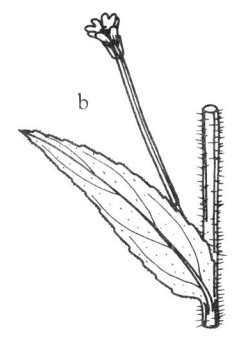

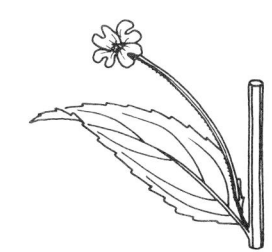

Flower and leaf of BROAD-LEAVED WILLOW HERB

WINTER HELIOTROPE
(Compositae)

If you put an 's' in the first syllable of the Latin name *Petasites fragrans*, this would better describe the pestiferous WINTER HELIOTROPE. Nobody can quarrel with the *fragrans*, for it is strongly sweet-smelling at a time of year when few other flowers are to be seen. I suppose that every gardener has at some time put up with it for this reason, not much later discovering that the agreeable patch has spread like the plague, the 4–8in (10–20cm) heart-shaped leaves overshadowing other plants, the white fleshy rhizomes tunnelling under them. The plant was discovered in France on Mont Pilat in 1806 and introduced into Britain as a welcome winter-flowerer. It became a popular pot plant in Victorian parlours and, put out of doors, was found to thrive equally well in the garden. It defies eradication. Leave a shred of root and it will be with you again. The leaves grow singly from the creeping roots, as do the long stems bearing clusters of soft maroon-and-white flowers from January to March. Being a Composite, the flower-head is composed of many florets. The hairy flower-stalk emerges from a sheath that remains on the stem like a leaf. Occasionally the stalk branches to produce another inflorescence. The Winter Heliotrope is now naturalized in scattered places throughout Great Britain and Ireland, common in the western Mediterranean regions.

A species found locally in the U.S. and Britain is *Petasites hybridus*, the BUTTERBUR, with leaves often 2ft (60cm) across.

WINTER HELIOTROPE

YARROW (Compositae)

Another weed not to be encouraged is the YARROW or MILFOIL, *Achillea millefolium*, because, again, its roots are far-creeping and soon produce a forest of stems varying in height between 3in and 2ft (8–60cm). '*Millefolium*' gives the clue to its alternate leaves which are lanceolate in outline and 2–3 times pinnate, so finely divided and redivided that they look like ruffled feathers. The plant contains what is called an ethereal oil and this is what gives it its aromatic scent. The heads of white flowers from June to September are in dense terminal clusters, each flower-head having usually 5 ray-florets surrounding the white or sometimes cream-coloured disk-florets. The flowers are sometimes a soft pink, but it is still the same species. The plant was used medicinally, the astringent action of its crushed leaves stopping the flow of blood. Achilles, from whom it gets its generic name, used it to staunch the wounds of Telephus. Yarrow is common throughout the British Isles and Europe, Canada where it is also called PIPER, and the United States excepting areas in southern Texas and the south-western states.

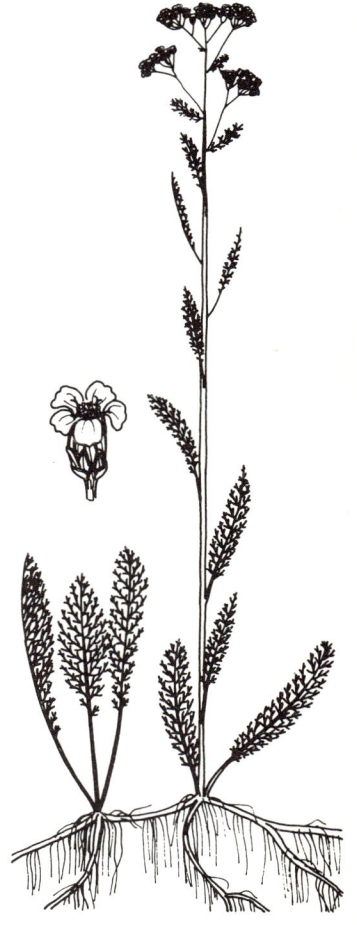

YARROW, with detail of flower-head

Glossary

Achene. Small dry single-seeded fruit.

Alkaloid. Naturally occurring basic nitrogenous substance, often poisonous.

Alternate. Describes a leaf arrangement on the stem when leaves arise alternately; cf. Opposite, Whorl.

Aphis, *pl.* **aphids.** Insect pests (e.g. Greenfly) attacking plants.

Apomixis. Reproducing by seed not formed from a sexual fusion.

Axil. Angle between leaf and stem.

Axillary. Arising in the axil of a leaf or bract.

Beak. Appendage at the tip of a fruit as in that of a buttercup, being the remains of the stigma and/or style.

Bifid. Split deeply in two. Cf. Trifid.

Bipinnate. *See* Pinnate.

Bract. Modified leaf beneath a flower or flower cluster, usually green but sometimes coloured, as in the Poinsettia, whose 'flower' is composed of bracts.

Bracteole. The one or two last bracts under each flower.

Bulbil. Small bulb arising in the aerial part of a plant.

Calyx. The sepals as a whole; cf. Corolla.

Capsule. Dry fruit splitting up when ripe.

Composite. Flower-head composed of disk-florets and/or ray-florets (e.g. Daisy, Dandelion).

Compound. Applied to a leaf, made up of several distinct leaflets; to a flower, a branching inflorescence.

Corolla. The petals as a whole; cf. Calyx.

Corymb. Inflorescence with all the flowers at approximately the same level, the outer flowers opening first.

Crenate. Describes a leaf-margin with roundish teeth.

Cyme. Type of flower-head on which the growing points end in a flower, the central flowers opening first.

Crisped. Curled.

Decumbent. Applied to a stem, lying on the ground and tending to rise at the end.

Diptera. Two-winged flies (e.g. midge).

Escape. A garden plant escaped into the wild.

Ethylene. An ether exhaled by some plants (e.g. Dandelion).

Family. Category of biological classification (e.g. Cruciferae) containing one or more genera (e.g. *Brassica*, *Sisymbrium*), each genus containing one or more species (e.g. *Brassica kaber*, *B. oleracea*).

Floret. A small flower, with or without sexual organs, many of which together form a single 'flower' (e.g. Daisy, Dandelion), disk-florets being in the centre, ray-florets being the strap-shaped petals.

Genus, *pl.* **genera.** *See* Family.

Herbicide. Weedkiller.

Hermaphrodite. Containing both male and female organs.

Hydroxyl. Chemical compound of hydrogen and oxygen.

Inflorescence. The arrangement of flowers on a plant.

Involucre. A series of bract-like leaves below the flower cluster.

Keel. Lower petal or petals shaped like the keel of a boat. *See* Papilionate.

Lanceolate. Lance-shaped.
Latex. A milky juice.
Linear. Long and narrow.
Lobed. Divided but not into separate leaflets.
Locally. Only in certain localities.

Margin. Edge of a leaf.
Mosaic. Types of virus diseases affecting plants.

Nematode. Eelworm.
Node. A point on the stem where one or more leaves arise.

Ochrea. Sheathing stipules of a leaf.
Opposite. Describes a leaf arrangement when two leaves arise at the same level on opposite sides of the stem; cf. Alternate, Whorl.
Ovary. The part of the female organ enclosing the ovules, which develop into seed after fertilization.

Palmate. Compound, consisting of more than three leaflets arising from the same point.
Panicle. Branched inflorescence, usually conical in outline. Cf. Raceme.
Papilionate. Flower of the pea family having an upper petal called the standard, two side petals known as wings, and two lower ones united into the keel. The term alludes to the fancied resemblance to a butterfly.
Perianth. A collective term for petals and sepals (corolla and calyx).
pH scale. Scale by which the degree of acidity or alkalinity in the soil is measured, below pH7 being acid; above, alkaline. pH is the symbol for 'hydrogen ion potential'.

Pinnate. Composed of more than three leaflets arranged in two rows along a common stalk. Bipinnate: in which these divisions are themselves pinnate. Similarly tripinnate.
Pistil. The female organ comprising (1) the stigma at the top, which receives the pollen; (2) the style, the tube down which the pollen grows to reach (3) the ovary at the bottom, and so fertilize the ovules.
Pollination. The act of depositing pollen from one flower on the stigma of the same or another flower, usually on a different plant of the same species.

Raceme. Unbranched inflorescence, the flowers being borne on short stalks.
Revolute. Rolled downwards.
Rhizome. Creeping underground stem (*adj.* rhizomatous).
Rhomboid. Having the shape of the diamond in a pack of playing cards.
Rootstock. Elongated underground stem.
Rosette. Cluster of leaves forming a circle on the ground, as with the Dandelion.

Segment. Division of a compound leaf or of a perianth.
Sepals. The outer segments of a flower, usually green or leaf-like, forming the calyx.
Sessile. Lacking a stalk.
Spadix. Thick column in such plants as the arum, formed by the male and female flowers growing together.
Spathe. Bract enclosing one or several flowers, in the arum forming a hood over the spadix.
Spatulate. Shaped like a paddle.
Species. *See* Family.
Spike. Elongated inflorescence with sessile or nearly sessile flowers.

Spikelet. Small spike; the ultimate flower cluster of the inflorescence of grasses and sedges.

Stamen. Male organ in the flower, producing pollen.

Standard. *See* Papilionate.

Stigma. *See* Pistil.

Stipule. Outgrowth at the base of a leaf-stalk in such plants as the vetch.

Stolon. Flexible runner in plants like the strawberry, which forms new roots at a distance from the parent plant (*adj.* stoloniferous).

Style. *See* Pistil.

Symbiosis. Beneficial relationship between one plant and another.

Tap-root. Vertical main root, bearing small fibrous roots.

Tendril. Slender coiling structure modified from a leaf, by which some plants climb; in vetches forming the terminal portion of the pinnate leaf.

Terminal. Borne at the end of a stem and limiting its growth.

Ternate leaf. Compound leaf consisting of three more or less equal leaflets.

Terpene. Substance present in such weeds as the Canadian Fleabane, acting as a skin irritant.

Testa. Skin or outer coat of a seed.

Trace elements. Literally traces of minerals in the soil, such as iron, zinc, copper, manganese and cobalt, necessary for the well-being of plants.

Trifid. Divided deeply into three. Cf. Bifid.

Tripinnate. *See* Pinnate.

Tuber. Thickened underground stem.

Umbel. Umbrella-shaped inflorescence where the stalks bearing the flowers arise from a single axis.

Whorl. Arrangement of leaves in a circle around the stem; cf. Alternate, Opposite.

Bibliography

CIBA–GEIGY *Weed Communities of Europe* 1971

CLAPHAM, A. R., TUTIN, T. G. AND WARBURG, E. F. *Flora of the British Isles* London and New York, Cambridge University Press, 2nd edition 1962

COCANNOUER, JOSEPH A. *Weeds: Guardians of the Soil* Old Greenwich, Ct, Devin-Adair, 1950

EWAN, JOSEPH AND NESTA *John Banister and His Natural History of Virginia, 1678–1692* Urbana, Il, and London, University of Illinois Press, 1970

FOGG, JOHN M., JR. *Weeds of Lawn and Garden* Philadelphia, University of Pennsylvania Press, 1945; reprinted New York, Hafner, 1956

FRIEND, HILDERIC *Flowers and Flower Lore* (2 vols) London, Swan Sonnenschein, 1883; reprinted Detroit, Gale, 1973

GRAY, ASA *Manual of the Botany of the Northern United States* Chicago, Ivison and Phinney, 1858; 8th edition New York, Van Nostrand Reinhold, 1950

HANF, MARTIN *Weeds and Their Seedlings* BASF United Kingdom Limited, 1972

HARWOOD, W. S. *New Creations in Plant Life: an authoritative account of the life and work of Luther Burbank* New York, Macmillan, 1907

HATFIELD, AUDREY WYNNE *How to Enjoy Your Weeds* London, Frederick Muller, 1973; New York, Macmillan, 1973

HEDRICK, ULYSSES PRENTISS *A History of Horticulture in America to 1860* New York, Oxford University Press, 1950

HUBBARD, C. E. *Grasses* Harmondsworth, Penguin, revised edition 1968

JAMES, W. O. *Background to Gardening* London, Allen & Unwin, 1957

JOSSELYN, JOHN *New England's Rarities Discovered ...* London, 1672

KING, LAWRENCE J. *Weeds of the World* London, Leonard Hill Books, 1966; New York, Wiley, 1966

LONG, H. C. *Weeds of Arable Land* (Ministry of Agriculture and Fisheries Bulletin No. 108) London, 1938

MONTGOMERY, F. H. *Weeds of Canada and the Northern United States* Toronto, The Ryerson Press, 1964; London, Bailey Brothers & Swinfen, 1964; New York, Frederick Warne, 1965

MOORE, F. JOAN AND THURSTON, JOAN M. 'Interrelationships of fungi, weeds, crops and herbicides' *Proc. 10th Weed Control Conference*, 1970

NORTH, PAMELA *Poisonous Plants and Fungi in Colour* London, Blandford, 1967

PHILBRICK, HELEN AND GREGG, RICHARD B. *Companion Plants* Old Greenwich, Ct, Devin-Adair, 1966; London, Stuart & Watkins, revised edition 1967

REID, CLEMENT *The Origin of the British Flora* London, Dulan and Co., 1899

RIDLEY, HENRY N. *The Dispersal of Plants throughout the World* Ashford, Kent, L. Reeve and Co. Ltd., 1930

ROHDE, ELEANOUR SINCLAIR *The Old English Herbals* London, Longman, 1922; reprinted New York, Dover Publications, 1971

SALISBURY, E. J. *The East Anglian Flora* Norfolk and Norwich Naturalists' Society, 1932
The Living Garden London, Bell, 1936
Weeds and Aliens London, Collins, 1961

SMITH, JAMES EDWARD *English Botany* London, 1832

SMITH, K. M. *A Textbook of Plant Virus Diseases* 2nd edition London, Longman, 1957; New York, Academic Press, 1972

SPENCER, EDWIN ROLLIN *All About Weeds* New York, Dover Publications, 1974

THURSTON, JOAN M. (compiled by) 'Some examples of weeds carrying pests and diseases of crops' *Proc. 10th Weed Control Conference*, 1970

UNITED STATES DEPARTMENT OF AGRICULTURE *Selected Weeds of the United States* (Agricultural Handbook No. 366) 1970

YOUNG, D. P. '*Oxalis* in the British Isles' *Watsonia*, **4**, 51–69 (1958)

Acknowledgements

It is to my many kind correspondents all over the British Isles and overseas in continental Europe and the United States of America that I am principally indebted for information about the weeds which are the subject of this book. Most obligingly they replied to a set list of questions, many of them supplying extra data that was of much help to me, and even placing themselves at my disposal as courts of appeal when weeds and their vagaries were in debate. In this way I was able to compile a master list of weeds common to all these countries, those common to Europe and Britain, and those native to Europe, to Britain, and to the United States. I therefore thank these garden owners and individual members of horticultural societies, directors of botanic gardens, research stations and municipal parks and gardens, professors of botany, county horticultural advisers and schools of horticulture, for the interest they expressed in my project, their encouragement, and their help so willingly given.

It was evident that the book I planned had a public waiting for it in 'suffering gardeners', for the seventy-five per cent response to my appeal was truly remarkable, and many were the comments revealing a more than casual interest in weeds. Some were passionate. 'Out of a very extensive collection,' wrote one correspondent, owner of a famous shrub garden, 'I place the most damnable in order as follows ...' There was a bonus from correspondents passing on my appeal to their gardening friends.

In particular I wish to thank those members of the staff of the National Botanic Gardens, Glasnevin, who supplied lists that between them covered a fair cross-section of Ireland; the National Trust and especially Paul Miles, the Trust's Horticulturist; the National Trust for Scotland; the Agricultural Research Service, United States Department of Agriculture; CIBA-Geigy for special information on European weeds; and for other specialized information A. Redcliffe-Smith of the Royal Botanic Gardens, Kew; Dr S. M. Walters and Dr Peter Yeo of the University Botanic Garden, Cambridge; William Lanier Hunt, the Hunt Arboretum; the Weed Research Organisation; the Soil Association; the Ministry of Agriculture and Fisheries; the Henry Doubleday Research Association; the Botanical Society of the British Isles and its member Eric J. Clement; Dr G. D. Heathcote, Broom's Barn Experimental Station; and the Rothamsted Experimental Station, particularly Miss J.M. Thurston. For the Table of Weedkillers on pages 44 and 45 and for information on selective weedkillers I thank the several scientific officers of Fisons Limited whose experiments I was able to watch, and others for added information. I also thank the libraries which assisted me with long-term loans of books, including the Suffolk County Library, the Lowestoft Library and the library of the John Innes Institute.

For botanical help during the writing of the book I am grateful to A. Gavin Brown, Kenneth Campbell and particularly John S. L. Gilmour and Victoria Matthews, to whom I am further grateful for her beautiful illustrations relating to the details of my text, and to Grace Woodbridge for her excellent photographs showing weed communities and specimens illustrating particular aspects of habit and habitat, whose help also in integrating text and illustrations was invaluable.

I hope the result will be of profit to all gardeners who read these pages, and, perhaps, in some cases, even to the weeds themselves.

Index of Botanical and Common Names

General Index